Experimental Entomology

Reinhold Books in the Biological Sciences

CONSULTING EDITOR: PROFESSOR PETER GRAY

Department of Biological Sciences
University of Pittsburgh
Pittsburgh, Pennsylvania

The Encyclopedia of the Biological Sciences, edited by Peter Gray

Biophysics: Concepts and Mechanisms, by E. J. Casey
Cell Function, by L. L. Langley
Chordate Morphology, by Malcolm Jollie
Concepts of Forest Entomology, by Kenneth Graham
Ecology of Inland Waters and Estuaries, by George K. Reid
Environmental Measurement and Interpretation, by Robert B. Platt and John F. Griffiths
Evolution: Process and Product, Revised Edition, by Edward O. Dodson
Experimental Biology: Measurement and Analysis, by R. H. Kay
General Zoology, by Clarence J. Goodnight, Marie L. Goodnight, and Peter Gray
Introduction to Comparative Entomology, by Richard M. Fox and Jean Walker Fox
Management of Artificial Lakes and Ponds, by George W. Bennett
Manual of Insect Morphology, by E. Melville DuPorte
Natural History, by Richard A. Pimentel
Paramedical Microbiology, by Stanley E. Wedberg
The Plant Community, by Herbert C. Hanson and Ethan D. Churchill
Principles in Mammalogy, by David E. Davis and Frank B. Golley

Consulting Editor's Statement

One of the difficulties in presenting a good laboratory course in entomology has always been the dearth of background material for other than taxonomic studies. The publication of Fox and Fox's "Introduction To Comparative Entomology" has pointed up this difficulty since the broad approach in this work renders even more difficult the design and selection of laboratory exercises.

I am well aware that no two courses are identical either in their approach or in the facilities available. However, the material in this new book should make it easier for the instructor to develop his own laboratory program. It presents a wealth of well-documented background material from which he can select and design a program as he wishes.

This volume was written in such a way that it may be used with the Fox and Fox book "Introduction To Comparative Entomology" or with any other well-known entomology text.

PETER GRAY

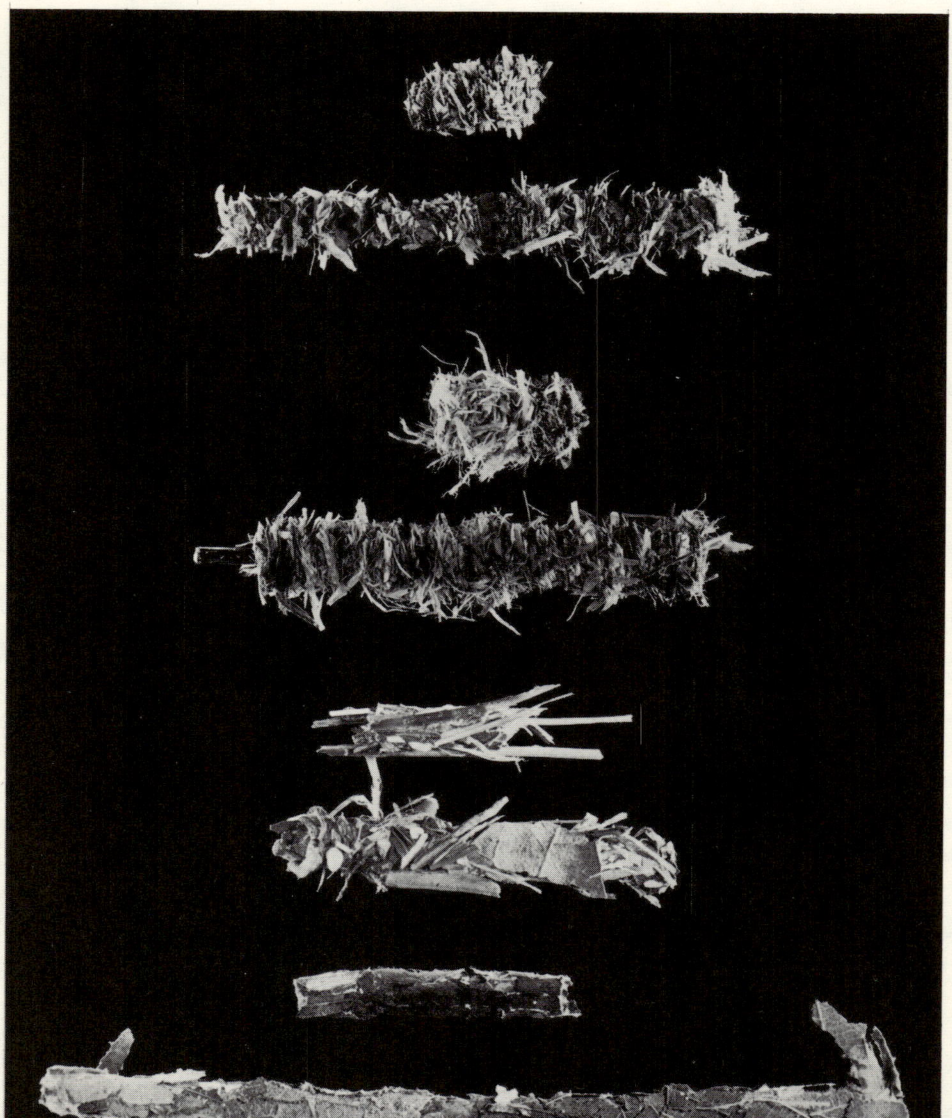

Abnormally elongated caddis fly larval cases produced experimentally, with normal (control) cases. Reading from top: first, *Limnephilus indivisus* Walker; second, *Platycentropus radiatus* (Say); third, *Pycnopsyche guttifer* (Walker); fourth, *Ptilostomis semifusciata* (Say). Photographs by courtesy of D. Merrill, University of Michigan.

Experimental Entomology

KENNETH W. CUMMINS
University of Pittsburgh
Pittsburgh, Pennsylvania

LEE D. MILLER
University of Pittsburgh
Pittsburgh, Pennsylvania

NED A. SMITH
University of Pittsburgh
Pittsburgh, Pennsylvania

RICHARD M. FOX
Carnegie Museum and
University of Pittsburgh

NEW YORK
REINHOLD PUBLISHING CORPORATION
CHAPMAN AND HALL LTD., LONDON

Library of Congress Catalog Card Number: 64-8548
Printed in the United States of America

PREFACE

A little more than a year ago Professor Cummins undertook to organize a laboratory program in general entomology to accompany my lectures at the University of Pittsburgh. From the outset, Professor Cummins and I wanted to apply to entomology some of the contemporary methods and approaches being used in other branches of biology with so much success. To this end we determined to emphasize the use of living insects and experimental techniques in the laboratory and to curtail the more traditional exercises so largely preoccupied with the structure and classification of dead insects. The resultant program was so successful and so stimulating to the students that we were asked to make it generally available in book form.

We at once enlisted as coauthors two colleagues in the Department of Biological Sciences who had contributed many fresh, useful ideas to the laboratory program: Lee D. Miller—taxonomist, morphologist and geneticist—and Ned A. Smith—insect physiologist and expert in experimental methods. It is now impossible to decide which of the four of us has contributed the most to this volume: it is truly the product of a close and happy collaboration.

This resultant book is intended to serve both as a laboratory guide and as an idea box. The first two chapters provide exercises in "plumbing, wiring and engineering" and in identification—presented as essential background for experimental work rather than as the major objective of the introductory course. The next four chapters detail, under the headings of genetics, physiology, behavior and ecology, a series of exercises in the solution of some fundamental biological questions through the manipulation and observation of living animals. An appendix provides information on techniques, recipes, culture methods and sources of supply.

We have included in this book far more work than a class can be expected to accomplish during a single term and we hope thereby to afford maximum flexibility to the professor by enabling him to select exercises in keeping with his own interests and the facilities available to him. We have avoided outlining a syllabus for the laboratory work in general entomology in the belief that any cut-and-dried scheme underrates both the student and the teacher. We assume that our teaching colleagues prefer (as we do) to develop their own programs and schedules and we have tried to prepare a book that will assist them in doing so.

Experimental Entomology is designed primarily for use in college and university courses in general entomology, but it should also prove useful as a guide for special projects and as an introduction to entomological research by individuals or classes at any level. While this book is intended as a companion volume and supplement to Fox and Fox, *Introduction to Comparative Entomology,* it can nevertheless be used with any standard textbook on the subject.

We are grateful to Professor Charles L. Ralph for reviewing the material pertaining to physiology and to Professor Eliot B. Spiess for reviewing the chapter on genetics. Professor C.K. Dorsey of West Virginia University read the entire manuscript and his criticisms and suggestions were exceedingly valuable.

Richard T. Satterwhite, Section of Insects and Spiders, Carnegie Museum, prepared all drawings and their excellence contributes, we believe, to the value of the book.

Mention must be made of the numerous suggestions and contributions made by our graduate students at the University of Pittsburgh.

We particularly appreciate the cooperation and assistance given us in the preparation of this book by editors Jim Ross, Len Roberts and Cynthia Harris of Reinhold. They helped us solve problems at all stages of production, without in any way interfering with our notions of what to include and how to write—one of the consequences of which is that the onus for errors rests solely with the authors.

October, 1964

R. M. Fox

CONTENTS

Preface vii

1. Morphology **1**

A. The Giant Cockroach, Blaberus 1
1. External morphology, *2.* 2. Internal morphology, *8.*
B. Caddis Fly Larvae 9
1. External morphology, *9.* 2. Internal morphology, *11.*
C. The Fruit Fly, Drosophila 13
1. External morphology, *13.*
D. The Flour Beetle, Tribolium 15
1. External morphology, *15.* 2. Distinguishing T. confusum from T. castaneum, *15.* 3. Separating males and females, *16.*
E. Genitalia of Lepidoptera 16
1. Procedure, *17.* 2. Observations, *18.*
F. Morphological Variations 19
1. Insects, *20.* 2. Myriapoda, *23.* 3. Arachnida, *23.*

2. Systematics **25**

A. Keys 26
1. Key to the classes and orders of terrestrial arthropods, *29.* 2. Key to the genera of Romaleinae of America north of Mexico, *33.*
B. Behavorial and Physiological Taxonomy 35
1. Behavioral identification, *35.* 2. Physiological identification, *48.*

3. Genetics **50**

A. Genetic Crosses with Drosophila 50
1. Single gene autosomal cross, *52.* 2. Single gene sex-linked cross, *53.*

B. Dipteran Giant Salivary Gland Chromosomes 54
C. Sex Determination in Habrobracon 56

4. Physiology **59**

A. Sensory Reception 59
1. Sexual attraction in saturniid moths, *60*. 2. Localization and nature of taste responses in adult Diptera, *62*. 3. Chemical factors determining food plant selection by lepidopterous larvae, *63*. 4. Temperature and humidity responses, *67*.
B. Blood 71
1. Hemocytes, *71*. 2. Qualitative analysis of free amino acids in hemolymph by paper chromatography, *72*.
C. Respiration 74
1. Oxygen consumption by terrestrial insects, *75*. 2. Oxygen consumption by aquatic insects, *78*. 3. Carbon dioxide production by aquatic insects and determination of RQ, *85*.
D. The Temperature Coefficient Q_{10} 88
1. Effect of temperature on oxygen consumption (QO_2), *89*. 2. Effect of temperature on insect heart rate, *89*. 3. Effect of temperature on dye transport by the malpighian tubules, *90*.
E. Neural Physiology 91
1. Metameric independence of the insect nervous system, *91*. 2. Spontaneous electrical activity of insect ganglia, *92*.
F. Hormonal Control of Insect Growth and Development 95
1. Extraction of juvenile hormone, *97*. 2. Wigglesworth method of bioassay, *98*. 3. Compounds having activity, *100*.

5. Behavior **101**

A. Caddis Fly Case-Building Behavior 101
1. Case-building patterns, *103*. 2. Ontogeny of case building, *108*.
B. Courtship Behavior of Drosophila 110
C. Communication by Stridulating Orthoptera 111
D. Web Building and Orientation in Spiders 113

6. Ecology **117**

A. Population Analysis 117
1. Field population census, *118*. 2. Population sampling

by chemoattraction of saturniid moths, *122.* 3. Population sampling with an ultraviolet light trap, *123.* 4. Laboratory population census, *124.* 5. Interspecies competition, *129.*

B. Community Trophic Dynamics and Energetics 131

1. Trophic relations of insects ingesting solid food, *133.* 2. Trophic relations of insects ingesting fluids, *135.* 3. Tracer techniques, *135.* 4. Energetics, *141.*

Appendix **143**

Live Materials and Cultures 143

Butterflies, *146.* Cockroaches, *146.* Drosophila, *147.* Gryllus, *147.* Habrobracon and Anagasta, *147.* Saturniidae, *148.* Spiders, *148.* Tenebrio, *149.* Tribolium, *149.* Trichoptera larvae, *149.*

Media and Formulation 150

Juvenile hormone, *150.* Lacto-aceto-orcein solution, *150.* Mountants, permanent, *150.* Mountants, temporary, *150.* Ninhydrin spray, *151.* Pringle's saline solution, *151.* Sarkarig's stain, *151.* Tap water conditioning, *151.* Whole mounts, *151.*

Special Equipment 152

Humidity and temperature measuring devices, *152.* Insect songs, *153.* Millipore filters, *153.* Nerve chamber, *153.* Oxygen bomb calorimetry, *153.* Oxygen electrode instruments, *153.* Tracer equipment, *153.*

Keys 154

References Cited **155**

Index **171**

Experimental Entomology

CHAPTER ONE

Morphology

A knowledge of structure is a necessary preliminary to the understanding of function and behavior. The exercises in this chapter are designed to afford a basic general familiarity with the morphology of the insects that will be used in the experiments outlined in later chapters.

The giant cockroach *Blaberus* has been selected to begin the series because it is among the more generalized and structurally primitive of living insects. We then turn to the caddis fly larva as an example of an immature insect which, despite its relatively unmodified body, has evolved a complex set of behavioral patterns with remarkable survival value. The fruit fly *Drosophila,* invaluable in genetic and population studies, has a highly evolved embryogenesis leading to adult structures well modified as compared to the primitive pattern. The beetle *Tribolium* is useful experimentally, but males and females of the several closely related, structurally similar species are difficult to distinguish. The genitalia of many insect groups, notably Lepidoptera, are of primary value in taxonomic work, and a section is included on genitalic structure.

This chapter concludes with suggestions on selecting other terrestrial arthropods, the examination of which would afford desirable breadth to the study of morphology.

A. THE GIANT COCKROACH, *BLABERUS*

The cockroach is easily cultured in the laboratory (see Appendix) and is an excellent animal to use for the demonstration of nervous functions in insects (Chapter Four). Morphologically, the cockroach is little changed from the primitive orthopteroids that flourished during Carboniferous time some 280 million years ago, and it affords a good base line for further study of insect morphology.

For laboratory examination of the cockroach, specimens freshly killed with ether are the easiest to use. For the study of sclerotized parts without interferences from soft tissues, the specimens should be wetted in alcohol and put into a hot solution of 10% KOH for about an hour (see below) or left overnight in a cold solution.

MATERIALS. *Blaberus giganteus* or similar cockroach; dissecting microscope; dissecting pan; set of dissecting instruments, especially forceps, fine scissors, needles and pins; ether killing bottles; 10% solution KOH; test tubes; watch glasses; Bunsen burner or alcohol lamp.

1. EXTERNAL MORPHOLOGY

From the dorsal aspect (Fig. 1.1B) the two most prominent features of the cockroach are the **pronotal shield** (PnS), covering the head and anterior portion of the thorax, and the **forewings (tegmina)** (Fw), covering the posterior part of the thorax and the abdomen.

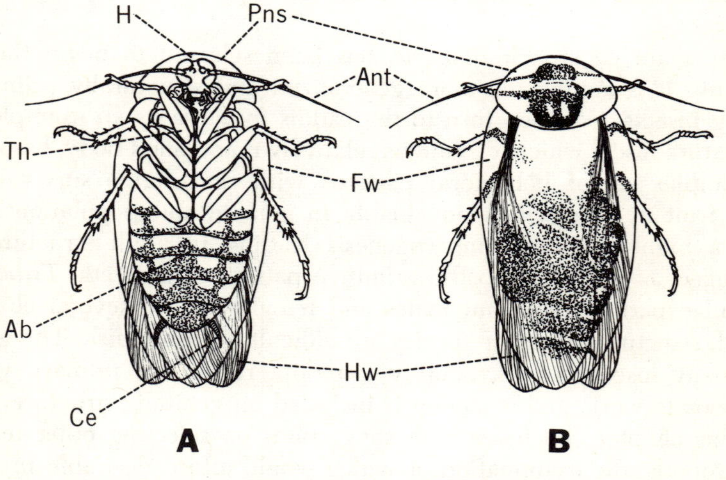

Figure 1.1. External morphology of *Blaberus giganteus.* (A) ventral view; (B) dorsal view. Ab, abdomen; Ant, antenna; Ce, cercus; Fw, forewing; H, head; Hw, hindwing; Pns, pronotal shield; Th, thorax.

From the ventral aspect (Fig. 1.1A) the **head** (H), **thorax** (Th) and **abdomen** (Ab) may be readily seen. These regions, common to all insects, and their appendages will be considered below.

HEAD. The head is attached to the first thoracic segment by a rather long, somewhat constricted membranous **cervix (neck)**. There are lateral sclerites on the cervix and some of these **cervical sclerites** are important taxonomically in some insect groups.

Note that the mouthparts are diverted backwards, an adaptation to the cockroach's mode of feeding. Such mouthparts are termed **hypognathous** (Fox and Fox, 1964: 50). Examples of other orientations of the mouthparts in relation to the head will be seen in other exercises.

Remove the head by severing at the cervix and examine under the low power of a dissecting microscope. Refer to Fox and Fox (1964: 49, Fig. 2.8) for the position and identification of structures discussed in this and the next three paragraphs. The top of the head is the **vertex** (Vx), divided by the **epicranial (coronal) suture** (EcSut) into two lateral **parietal** areas. The **frons** (F) is bounded by the **antennae** and the anterior mouthparts. The **gena** (Ge) is the cheek region below the eye and above the lateral mouthparts, separated from the frons by the **genal suture** (GeSut) and from the mouthparts by the **subgenal suture.** The ring sclerites around the posterior part of the head are the **occiput** (Occ), separated from the vertex by the **occipital suture** (OccSut), and the **postocciput** (Pocc), divided from the occiput by the **postoccipital suture** (PoccSut).

The most prominent appendages of the head are the long, slender, many-segmented and filiform **antennae.** Note particularly the longer basal (proximal) segment, the **scape,** and the shorter bulbous second segment, the **pedicel.** The pedicel contains the sensory **Johnston's organ** (Fox and Fox, 1964: 194, Fig. 6.5C), which will not be visible. The remaining segments of the antenna form the **flagellum (clavola** or **shaft).** The antennae are set in a membranous **antennal foramen** bounded by the **antennal sclerite** (AntScl) and separated from the rest of the head by the **antennal suture** (AntSut). The **compound eyes** (Eye) are lateral of the antennae and almost completely surround them.

The compound eyes have a granular appearance. Examine them under the highest power of the dissecting microscope; the granulations should now appear hexagonal. Each hexagon is a **facet** and covers an **ommatidium** (plural, **ommatidia**), the visual unit of the compound eye. Mediad of the antennae are paired opaque white membranous spots, the **fenestrae** (singular, **fenestra**), the degenerate **lateral ocelli** (LatOc).

Mouthparts. The ventral portion of the frons is more precisely known as the **clypeus** (Cp). In most insects the frons and clypeus are separated by the **fronto-clypeal (epistomal) suture;** but this suture is absent in the cockroach and the entire area is perhaps better called the **fronto-clypeus** in this insect. Ventrad of the fronto-clypeus is hinged a heavy median flap, the **labrum** (Lr), which functions as the upper lip.

Remove the labrum and expose the paired lateral horny **mandibles** (Md), the main chewing structures (Fig. 1.2A). Remove the mandible carefully and note the heavily sclerotized toothed cutting edges and the two points of articulation with the head, the **anterior condyle (ginglymus)**

(AtCnd) and the **posterior condyle** (PtCnd) located at the proximal corners of the mandible.

Posterior of the mandibles are the paired segmented **maxillae** (singular, **maxilla**), shown in Figure 1.2B and C. The basal portion of the maxilla is the **coxopodite,** composed of a short proximal **cardo** (Cr) and the more distal **stipes** (Si). At the distal end of the stipes are the partially fused medial **laciniae** (singular, **lacinia**) (Ln) and lateral **galeae** (singular, **galea**) (Ga). The **maxillary palpus** (Mxp) arises from the distal lateral margin of the stipes and is divided into several segments, apparently homologous to the segments of a walking leg, and is the **telopodite.** The two short proximal segments are probably the **trochanters,** and the three longer terminal segments are probably homologous with the femur, tibia and tarsus.

Behind the maxillae is the **labium** (Li), serving as the lower lip, a complex structure formed by the fusion of the paired **second maxillae.** The dissected labium is shown in Figure 1.2D. Like the maxilla, the labium is composed of segmental elements homologous with the parts of a walking leg. Remove the labium carefully and examine it under the dissecting microscope. The large flat proximal segments is the **postmentum** (Ptm), sometimes called the **postlabium;** all the other segments form the **prelabium.** Attached proximally to the postmentum is the **prementum** (Prm), beyond which is the **ligula,** formed by the fusion of the medial **glossae** (singular, **glossa**) (Gs) and the lateral **paraglossae** (Pgs), corresponding respectively to the laciniae and galeae of the maxillae. The three-segmented **labial palpi** (Lip) arise from the lateral margin of the distal end of the prementum. The palpus, homologous with the telopodite as in the maxillae, probably consists of a proximal trochanter, a femur, and a distal tibia (perhaps fused with the tarsus).

On the midventral surface of the head, lying between all the sclerotized mouthparts, is the membranous cone-shaped **hypopharynx** (Hf) (Fig. 1.2B), which approximately delimits the **mouth.**

Thorax. In all insects the thorax is three-segmented, consisting of the anterior **prothorax,** the **mesothorax** and the posterior **metathorax.** All three segments bear walking legs, but only the meso- and metathorax bear wings, as is seen in the cockroach.

Refer to the figure of the sclerites of the generalized thoracic segment in Fox and Fox (1964: 52, Fig. 2.10); note that the pattern of these sclerites in the cockroach is much modified as a result of the dorso-ventral flattening of the body. Remove the wings and save them for examination later. A **notum (tergite)** (DScl) extends across the entire dorsum of each segment, but the most prominent of these in the adult cockroach is that of the prothorax, the **pronotal shield** (Fig. 1.1, Pns). The **pleural**

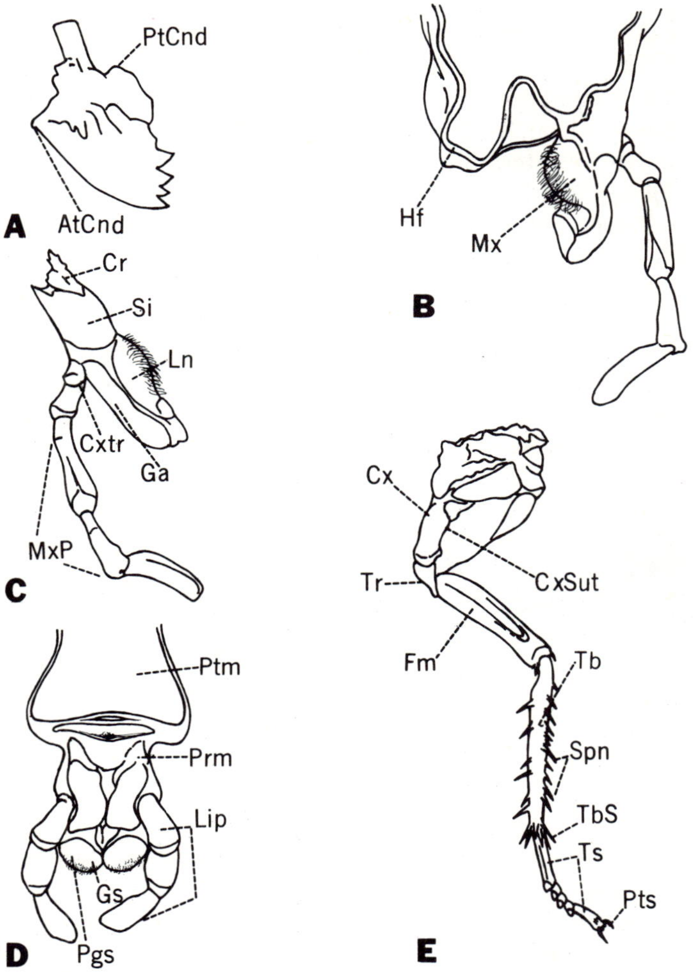

Figure 1.2. Morphologic details of *Blaberus giganteus.* (A) mandible; (B) maxilla and hypopharynx; (C) maxilla; (D) labium; (E) foreleg. AtCnd, anterior condyle; Cr, cardo; Cx, coxa; CxSut, coxal suture; Cxtr, coxotrochantral joint; Fm, femur; Ga, galea; Gs, glossa; Hf, hypopharynx; Lip, labial palpus; Ln, lacinia; Mx, maxilla; Mxp, maxillary palpus; Pgs, paraglossa; Prm, prementum; PtCnd, posterior condyle; Ptm, postmentum; Pts, post-tarsus; Si, stipes; Spn, spine; Tb, tibia; TbS, tibial spur; Tr, trochanter; Ts, tarsus.

sclerites lie on the ventral surface between the notum and the articulations of the legs. Note the transverse **pleural suture** (PlSut), which divides the pleural area into an anterior **episternum** (Epst) and a posterior **epimeron** (Epm). The episternal area is, in turn, divided again into an anterior **pre-episternum** (Pepst) and an **episternum** proper. The midventral plates, partially obscured by the legs, are the anterior, larger **sternum** (Vscl) and a smaller, posterior **sternellum** (Stl). The other sets of sclerites are present only on the two wing-bearing segments. Just anterior of the episternum is a small sclerotized ring, the **spiracle** (Sp), surrounding the external opening of the tracheal system. A prothoracic spiracle develops in insect embryos but is not present in adults (Fox and Fox, 1964: 51). Near the bases of the wings are the small inconspicuous **axillary sclerites** (AScl); some of these sclerites may have remained attached to the wing bases when they were removed.

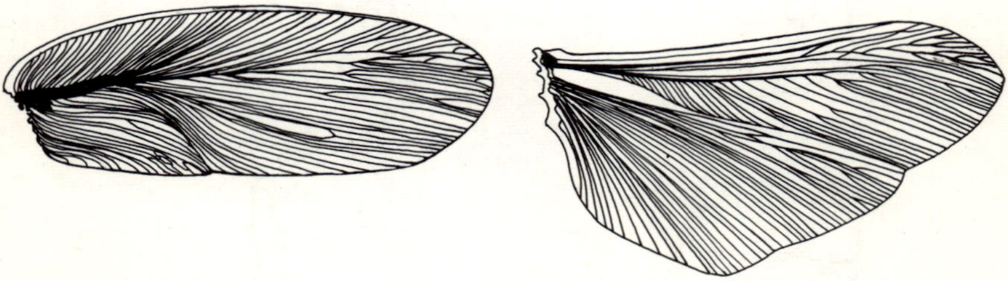

Figure 1.3. Forewing (left) and hindwing (right) of *Blaberus giganteus.*

WINGS. As shown in Figure 1.3, the venation of the cockroach wings is chaotic, presenting many difficulties in determining the homologies of the veins, the discussion of which is beyond the scope of this examination. An example of the hypothetical plan of venation is given by Fox and Fox (1964: 117, Fig. 4.3) and some of the major venational series in the cockroach wing can be extrapolated by comparison. The **forewings (primaries)**, being leathery and tough but not horny, are called **tegmina,** and the **hindwings (secondaries)** are membranous. It may be observed in the living cockroach that the hindwings are folded beneath the tegmina when at rest and are only displayed during flight.

LEGS. Remove one of the walking legs and examine it under the dissecting microscope. For comparison, see Figure 1.2E. The proximal segment, which articulates with the body, is the **coxa** (Cx). This represents only the distal portion of the original, prototypic coxa; the proximal parts are incorporated into the thoracic segment wall as pleural sclerites. The suture dividing the coxa longitudinally is the **coxal suture** (CxSut). The small **trochanter** (Tr) is attached to the distal end of the coxa and freely articulates with it by the **anterior** and **posterior trochantral condyles.** The trochanter is more or less immovably attached to the long **femur** (Fm), which in turn articulates with the heavily spined **tibia** (Tb). On the tibia note particularly the difference between the **tibial spines** (Spn) and the longer, terminal and partially movable **tibial spurs** (TbS). In many groups the presence or absence of such spurs is important for identification. The **tarsus** (Ts) is composed of five subsegments, the primitive number found in insects. Each tarsal segment bears gripping setae, an adaptation for clinging. At the tip of the distal tarsal segment, represented by the "tarsal claws" and the median gripping pad (**pulvillus**), is the tiny **post-tarsus** (Pts).

All leg elements are present and relatively unmodified in the cockroach, with the exception of the second trochanter. The trochanter found on the leg of *Blaberus* was probably formed by fusion of the first and second trochanters of the hypothetical ancestor. Only the Odonata, among all the living insects, possess two separate trochanters.

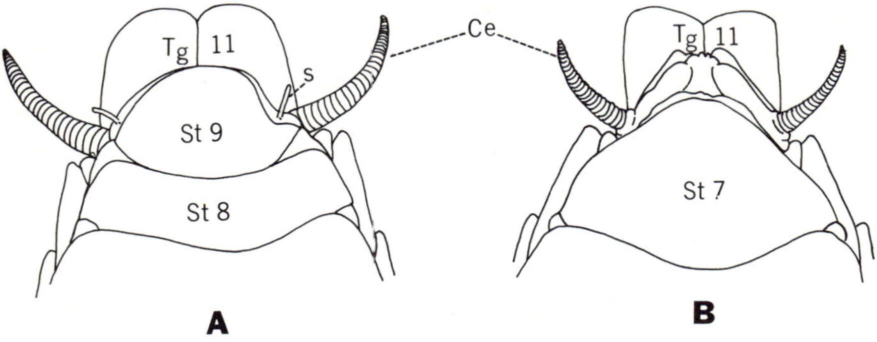

Figure 1.4. (A) male and (B) female abdominal termini of *Blaberus giganteus.* Ce, cerci; s, stylus; St 7, 8, 9, sternites of 7th, 8th and 9th abdominal segments; Tg 11, tergite of 11th abdominal segment.

Abdomen. The abdomen has eleven segments, not all of which are clearly visible in either sex. The first seven segments are always distinct, with separate **terga** (singular, **tergum**) and **sterna** (singular, **sternum**). On the first eight segments the spiracles lie between the dorsal and ventral elements. These anterior seven segments are generally unmodified, but the last four are highly modified and must be considered separately for the sexes.

In the *female* (Fig. 1.4B) the eighth and ninth sterna are hidden beneath the seventh (St 7), which is prolonged posteriorly into a hinged keel-like process. On the dorsal surface the eighth tergum is normal in appearance, while the ninth is much reduced and apparent only medially.

The tenth and eleventh (Tg 11) are fused into a smooth, laterally bilobate plate. The **cerci** (singular, **cercus**) (Ce) are appendages of the eleventh segment and extend laterally from its base. The cerci probably are homologous with the walking legs, but it is impossible to assign homologies to their parts.

In the *male* (Fig. 1.4A) both the sterna and terga of segments eight and nine are distinct; the tenth sternum is obscured by the ninth. The eleventh sternum is bilobate and flanks the anus; the two parts are known as the **podical plates.** The terga of segments ten and eleven (Tg 11) are fused. The males also bear cerci, and there are two other appendages, the little unsegmented **styli** (singular, **stylus**) (s), posteriorly diverted appendages extending from the ninth sternum.

2. INTERNAL MORPHOLOGY

The only internal structures to be detailed at present in the cockroach are in the head. However, in the section on neural physiology in Chapter Four a diagram of the central nervous system is given and it may be used as a guide to dissect the central nervous system of *Blaberus* at this point for comparison with that of the caddis fly larva, to be detailed below.

TENTORIUM. The cephalic apodemes, collectively known as the **tentorium,** serve as internal points of attachment for the mouthpart muscles; in the cockroach these structures do not differ significantly from those shown by Fox and Fox (1964: 50, Fig. 2.9).

Drop the cockroach head into a syracuse watch glass partially filled with alcohol. While the head is soaking, fill a test tube to a depth of about one inch with 10% KOH solution and, using a test tube holder, warm it over a Bunsen burner or alcohol lamp. (*Warning: agitate the test tube throughout the warming process in order to avoid having the KOH solution pop out of the tube. KOH is caustic and damages clothing as well as skin and eyes. Aim the test tube away from yourself and other people.*) When the solution is hot, drop the head of the cockroach into it and place the tube on a rack, letting it stand for about an hour. Then place the head in a syracuse dish of water to which a few drops of alcohol have been added to break the meniscus. Dissect as follows: with forceps pick away the integument of the vertex to about the level of the middle of the compound eyes; with a pipet flush out the extraneous soft material within the head. Observe the tentorium and compare it with Figure 2.9 in Fox and Fox (1964: 50).

The **anterior tentorial pits,** seen on external examination, are invaginations at the fronto-clypeal region forming the **anterior tentorial arms** (AtTt). The **dorsal tentorial arms** (DTt) arise near the front margin of the compound eyes but not from a definite pit. The **posterior tentorial arms** (PtTt) invaginate from the postoccipital suture seen previously. Connecting the right and left tentoria is a cross member, the **corporotentorium (tentorial bridge)** (Ctt).

The apodemes of the thorax may be dissected in the cockroach by using the KOH technique outlined above, but these structures are better seen in other insects. The plan of the thoracic apodemes is shown by Fox and Fox (1964: 53, Fig. 2.11).

B. CADDIS FLY LARVAE

The descriptions which follow apply specifically to larvae of the families Phyryganeidae and Limnephilidae (see Appendix), although with slight modifications they can be utilized for studies of any other Trichoptera except the Hydroptilidae (Nielsen, 1948). Variations among families particularly involve differences in the number and positions of sclerotized plates on the meso- and metathorax, modifications of the anal prolegs and hooks, and of the thoracic legs. In many species the ninth and tenth abdominal segments are fused.

A number of general descriptions of the morphology of trichopteran larvae have appeared in the literature (*e.g.*, Vorhies, 1905; Lloyd, 1921; Betten, 1934; Wessenberg-Lund, 1943; Ross, 1944; and Bullough, 1951). An excellent treatment of the external and internal anatomy, pertaining specifically to three limnephilids (*Platycentropus radiatus, Limnephilus indivisus,* and *Pycnopsyche guttifer*) and one phryganeid (*Ptilostomis semifasciata*), has been given by Merrill (1964).

1. EXTERNAL MORPHOLOGY

The structures discussed below should be observed in specimens preserved in a mixture of ten parts of 70% ethyl alcohol and one part of glycerin.

The general external morphology of a trichopteran larva is shown in lateral view in Figure 1.5A. This phryganeid has only the **pronotal shield** (PrS), whereas limnephilid larvae also have a mesonotum and scattered metanotal plates. The thoracic legs are well developed and there are a pair of **caudal prolegs** (Prl) which terminate in the **anal hooks** (Ah). These hooks, shown in greater detail in Figure 1.5B, C and D, are directed laterad and serve to hold the larva in its case. The importance of the sensory receptors associated with the hooks is discussed further in Chapter Five.

Additional detail of head and leg structure is shown in Figure 1.5E

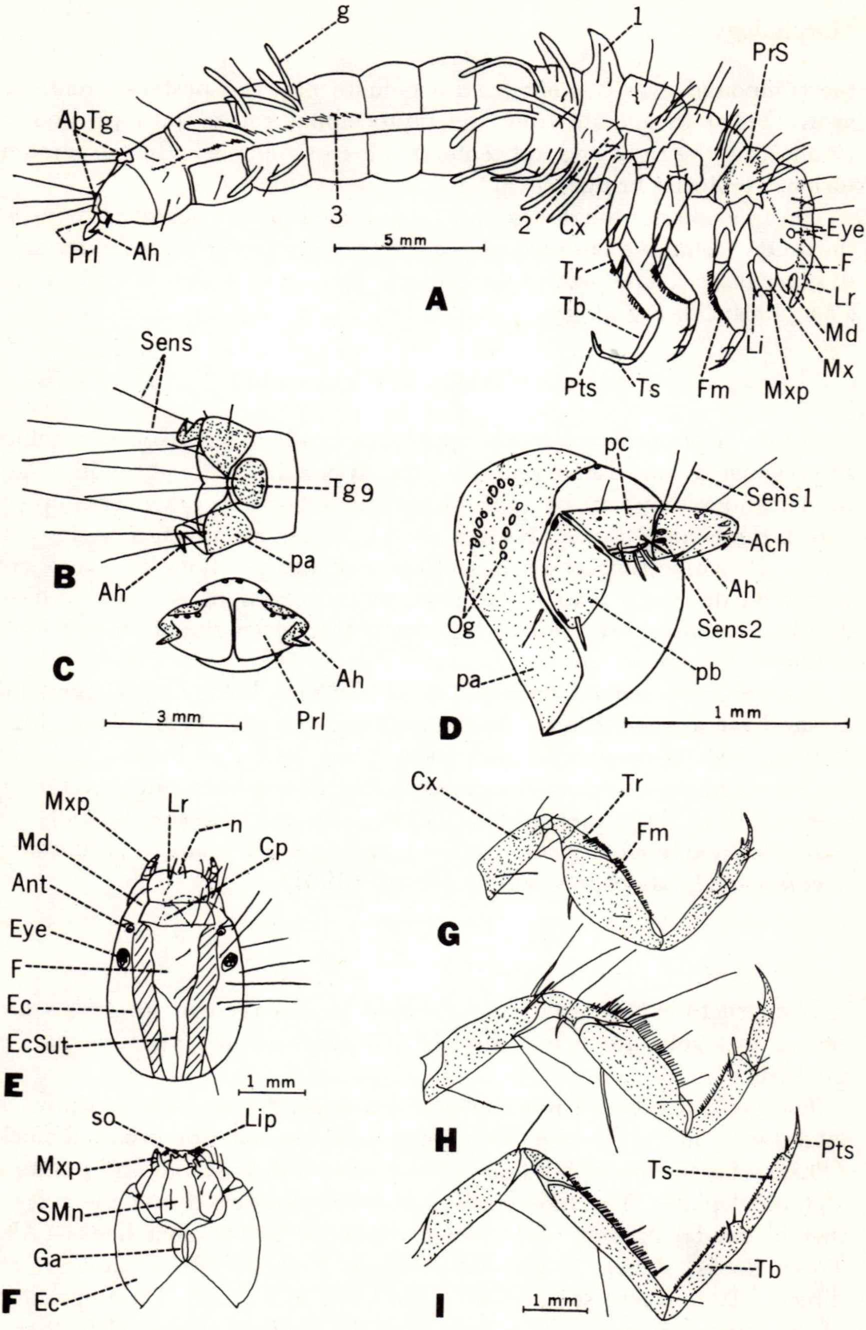

Figure 1.5. External morphology of a trichopteran larva, *Ptilostomis semifasciata* (Say). (A) lateral view of whole animal; (B) dorsal view and, (C)

and F. Note that the **labial palpi** (Lip) are reduced. The **silk orifice** (so) is the external opening of the **silk glands** (sGl) shown in Figure 1.6. The **labrum** (Lr) has a **medial notch** (n) rimmed with sensory receptors and is important in case-building behavior, as detailed in Chapter Five (see also Tindall, 1960). The only photoreceptors present in trichopteran larvae are small clusters of facets (Eye) representing the adult **compound eye.** The **antennae** (Ant) are greatly reduced (except in the Leptoceridae) and situated laterally between the eye facets and the bases of the **mandibles** (Md).

The head structures shown in Figure 1.5 should be compared with the generalized insect head given in Fox and Fox (1964: 43, Fig. 2.6; 49 Fig. 2.8). Also compare the larval legs of Trichoptera (Fig. 1.5G, H, I) with the above authors' figures of the generalized insect leg (Fox and Fox, 1964: 63, Fig. 3.1) and modifications of the basic plan (Fox and Fox, 1964: 74–75, Figs. 3.6 and 3.7). A useful comparison of the range of variation in external larval structures of Trichoptera is to be found in Betten (1934).

2. INTERNAL MORPHOLOGY

The larva should be pinned down in a cork- or wax-lined dissecting tray and slit longitudinally with a scalpel just lateral of the mid-dorsal line. A pair of large, coiled, whitish **silk glands** (sGl) (**spinning glands**) are readily seen (Fig. 1.6A); loops of these glands extend at least to the midabdominal region. The secreting portions of the glands are located in the thorax and abdomen. The tubule portions in the head region are concerned only with conducting the secretory product (Vorhies, 1905). In

---

Figure 1.5 (***continued***)

ventral view of 9th and 10th abdominal segments; (D) lateral view of left anal hook and the articulating plates of 10th abdominal segment; (E) dorsal view and (F) ventral view of head; (G) prothoracic leg; (H) mesothoracic leg; (I) metathoracic leg. AbTg, abdominal tergite; Ach, accessory hook; Ah, anal hook; Ant, antenna; Cp, clypeus; Cx, coxa; Ec, epicranium; EcSut, epicranial suture; Eye, eye; F, frons; Fm, femur; g, gill; Ga, galea; Li, labium; Lip, labial palpus; Lr, labrum; Md, mandible; Mx, maxilla; Mxp, maxillary palpus; n, notch; Og, origin of muscle; pa, pb, pc, plates a, b and c; Prl, proleg; PrS, pronotal shield; Pts, post-tarsus; Sens, sensory setae; Sens 1, trichoid sensilla; Sens 2, campaniform sensillum; SMn, submentum; so, silk orifice; Tb, tibia; Tg 9, tergite of 9th abdominal segment; Tr, trochanter; Ts, tarsus; 1, dorsal spacing hump; 2, lateral spacing hump; 3, lateral hair fringe. (Redrawn from Merrill, 1964, by courtesy of the author) the anterior region of the head the duct sections of the paired glands join to form a single tube opening to the exterior through the **silk orifice** (so), a small projection at the middle of the labium (Fig. 1.5F).

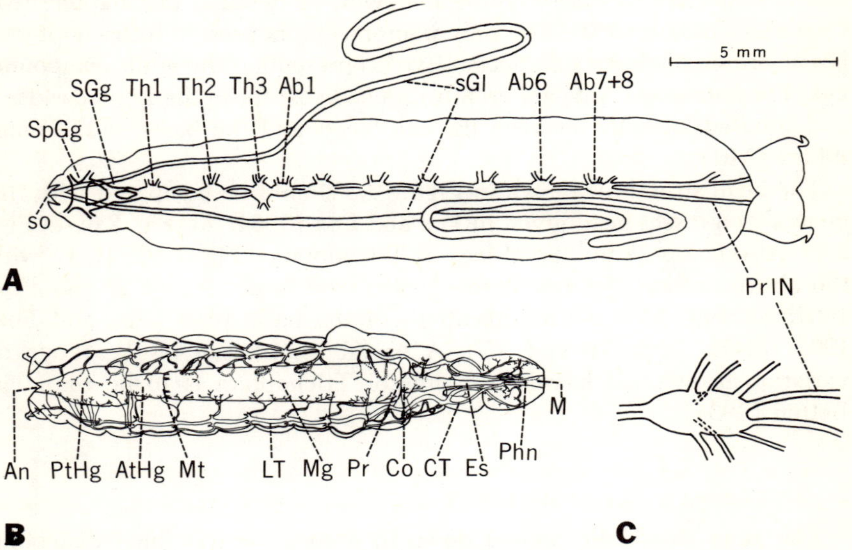

Figure 1.6. Internal morphology of a trichopteran larva, *Ptilostomis semifasciata* (Say). (A) dissection in dorsal view, the gut removed; (B) enlarged dorsal view of fused ganglia of 7th and 8th abdominal segments; (C) dorsal view of trachial and digestive systems, the malpighian tubules removed (not to scale of A). Ab 1, Ab 6, ganglia of 1st and 6th abdominal segments; Ab 7 + 8, fused ganglia of 7th and 8th abdominal segments; An, anus; AtHg, anterior hindgut; Co, crop; CT, cephalic trachea; Es, esophagus; LT, longitudinal trachea; M, mouth; Mg, midgut; Mt, malpighian tubules; Phn, pharynx; Pr, proventriculus; PrlN, nerve of proleg; PtHg, posterior hindgut; SGg, subesophageal ganglionic mass; sGl, silk gland; so, silk orifice; SpGg, supraesophageal ganglionic mass; Th 1, Th 2, Th 3, ganglia of the 1st, 2nd and 3rd thoracic segments. (A and B redrawn from Merrill, 1964, by courtesy of the author; C redrawn from Vorhies, 1904)

Removal of the silk glands reveals the digestive tract and the intricately branched tracheal system (Fig. 1.6C). There are two main **longitudinal tracheae** (LT) running the length of the body and a secondary system of branches results in a ring-like arrangement in each segment. The main tracheal trunks bifurcate in the head region into **dorsal** and **ventral cephalic tracheae** (CT). Small tracheal branches extend into the **external gills** in gill-bearing species. In addition, a highly intricate system of branches connects the main trunks with the various muscles and organs of the body.

The digestive tract (Fig. 1.6B) consists of a well-marked **foregut,** composed of the **pharynx** (Phn), **esophagus** (Es), **crop** (Co) and **proventriculus** (Pr), a long **midgut** (Mg), and a **hindgut** divided into **anterior** (AtHg) and **posterior** (PtHg) sections. Compare the caddis fly larvae gut with Figure 5.1 in Fox and Fox (1964: 136).

When the digestive tract is removed, the **ventral nerve cord** is exposed and easily seen (Fig. 1.6A, B). The **supraesophageal ganglion** (SpGg) can be exposed by careful dissection of the head capsule. Note in particular the positions of the thoracic and abdominal ganglia; similar dissections on living specimens will be necessary for physiological work outlined in Chapter Four. Compare the structures of the trichopteran nervous system with related figures in Fox and Fox (1964: 183, Fig. 6.3; 188, Fig. 6.4).

C. THE FRUIT FLY, *DROSOPHILA*

Since *Drosophila* species (see Appendix) will be used extensively in other experiments, some familiarity with the external morphology of these insects is essential. *Drosophila* may be taken further as typical of the Diptera, a group which varies widely from the generalized insect typified by the cockroach. Some information on the morphology of all stages of *Drosophila* is given by Strickberger (1962: 20–24) and Demerec and Kaufmann (1962: 7–9). Reference is made here to the structures and coloration of *D. melanogaster* (Meigen), but any species of *Drosophila* may be used with appropriate minor modifications of the descriptions below.

MATERIALS. Adult *Drosophila* (etherized wild-type flies), watch glasses, dissecting microscopes, fine needles for turning flies.

1. EXTERNAL MORPHOLOGY

Note again the three body regions: the head, thorax and abdomen. Unless otherwise indicated the structures mentioned below are illustrated by Demerec and Kaufmann (1962: 9), and these figures should be consulted as necessary as the exercise proceeds.

HEAD. The most prominent features are the bright red **compound eyes** (not specifically labeled). On the **vertex** note the small white **median** and **lateral ocelli** forming a triangle between the compound eyes; ocelli are absent in the cockroach but present in most insects. The **antennae** of *Drosophila* and many other Diptera are aberrant: there are three bulbous proximal segments with a series of feathery ones (the **arista**) arising from the middle of the third inflated segment. The **mouthparts** are highly modified and are known collectively as the **proboscis.** Such mouthparts are illustrated by Fox and Fox (1964: 94, Fig. 3.16D) and discussed by them in detail on page 95. The only structures readily recognizable in the mouthparts of *Drosophila* are the **maxillary palpi.** The **bristles** of the head, shown in detail by Demerec and Kaufmann, are of particular importance to the geneticist, but need be noted here only in passing.

THORAX. From the dorsal aspect only the **mesonotum** and its **scutellum** are visible; the **pronotum** is fused with the mesonotum, and the **metanotum** is hidden beneath the scutellum and may be seen only from the lateral aspect. From the side the **sternal** and **pleural** elements may be seen between the mesonotum and the legs. Identify these elements from the Demerec and Kaufmann figure (the terminology used is consistently applied in genetic literature, though it differs in some respects from that used in entomology texts). Only the **forewings** are developed in the Diptera. Note the reduction of **veins** as compared with the cockroach wing. In only a few insects, notably the hymenopterous superfamily Chalcoidea, is there a greater reduction in the number of veins. The **hindwings,** of *Drosophila* and other Diptera are modified into knobby, three-segmented structures called **halteres,** arising from the middle of the metanotum, and are organs of balance used during flight. In the **mesopleuron (pteropleuron),** the **axillary sclerites** may be distinguished under high power of the dissecting microscope. The legs of *Drosophila* are like those of the generalized insect with one exception: on the distal part of the first tarsal segment of the *male* foreleg is the prominent black **sex comb** which readily serves to distinguish the sexes. Again, attention is called to the **thoracic bristles** named in the figures of Demerec and Kaufmann.

ABDOMEN. In both sexes the **abdominal tergites** may be seen from the dorsal aspect. From the ventral aspect note the **abdominal spiracles** at the lateral edges of the tergites. The **abdominal sternites** are present on the ventral surface as oval, lightly sclerotized pads, each ornamented with many bristles.

Males may be distinguished by their relatively short, blunt, black-tipped abdomens (in wild *D. melanogaster*), whereas the female has a longer, more pointed abdomen banded yellow and black. The external genitalia of the two sexes are, of course, very different. Those of the female consist of a bifid **anal plate** at the distal end of the abdomen and a **vaginal plate** between the anal plate and the seventh sternite. The **ovipositor** opens between the vaginal and anal plates. The external genitalia of the male consist of a bifid **anal plate** in the corresponding position to that of the female, two lateral **genital arches** and a median **penis** positioned between the proximal ends of the genital arches. Homologies of these male and female genitalic structures may be derived from Fox and Fox (1964: 105–109).

D. THE FLOUR BEETLE, *TRIBOLIUM*

Tribolium species (see Appendix) will be used later in a competition experiment (Chapter Six), and familiarity with the basic structure of these beetles is desirable. In addition, the characters separating *T. confusum* (duVal) and *T. castaneum* (Herbst) and those distinguishing males from females must be considered.

MATERIALS. Adult *Tribolium*, preserved in alcohol; watch glasses; fine probes; dissecting microscopes.

1. EXTERNAL MORPHOLOGY

Place adult beetles in watch glasses and first observe the upper surface. From the dorsal aspect the **head** and **pronotum** are very prominent, the **meso-** and **metanota** and the **abdomen** being covered by the **elytra** (singular, **elytron**), the modified forewings. Between the elytra and the body are the hindwings, used solely for flight. Note the **striae,** the finely striated grooves on the elytra, important in recognizing the sexes of these beetles.

From the ventral aspect note the typical chewing mouthparts; like those of the cockroach they are **hypognathous,** or diverted downward. Locate the **labrum, mandibles, maxillae,** and **labium** as was done with the cockroach. From this aspect the **antennae** are also visible; note the general configuration of the antennae, since they are most reliable in separating *confusum* from *castaneum.* The three pairs of legs and the sternal elements of the three thoracic segments are all visible from the ventral aspect. Note also the abdominal **sternites** and **tergites** which were hidden from view by the elytra in the dorsal view.

2. DISTINGUISHING *T. confusum* FROM *T. castaneum*

Pfadt (1962: 489) separates *confusum* from *castaneum* by the larger size and smaller eyes of the former, in addition to the following two more reliable characters: the antennae are gradually enlarged distally in *confusum* (Fig. 1.7A), whereas the distal three segments of the antennae of *castaneum* (Fig. 1.7B) are abruptly enlarged and form a thick bulbous club. From the ventral aspect of the head, each eye of *confusum* is about one-third as wide, but the eyes of *castaneum* are about the same width as the distance separating them. *T. confusum* is flightless, but *castaneum* is capable of flight.

A **B**

Figure 1.7. (A) female of *Tribolium confusum.* (B) male of *T. castaneum.*

3. SEPARATING MALES AND FEMALES

The most reliable character, other than the configuration of the genitalia, for separating the sexes is found on the elytra. The posterior part of the sixth strial groove of the elytron, counting from the midline, in the female (Fig. 1.7A) swings mesad with the curvature of the elytron and unites with the third strial groove. The sixth strial groove of the male (Fig. 1.7B) is, in contrast, straight and ends at the margin of the elytron, not meeting the third strial groove (Pfadt, 1962: 489).

E. GENITALIA OF LEPIDOPTERA

Although the techniques listed below utilize Lepidoptera, they may be used with modifications for any group of insects. The most critical step is the treatment in KOH: if the abdomen is left too long in the solution, the sclerotized structures themselves may be affected in addition to the soft tissues. In general, it is better to "undercook" the structures. The terminology of the genitalia is that of Fox and Fox (1964: 105–109). The most comprehensive survey of the genitalia of all groups of insects is that edited by Tuxen (1956). The male and, to a lesser extent, the female genitalia are extremely important taxonomically.

MATERIALS. Pinned specimens of butterflies, preferably one of the larger species like the monarch, *Danaus plexippus* (Linné) (see Appendix); syracuse watch glasses—the standard 50 mm. (inside diameter) or, as some prefer, the smaller 20 mm. size; dissecting microscope; test tubes and test tube rack; test tube holder; alcohol lamp; dissecting needles

—some bent, others with straight tips; chips of glass (pieces from smashed slides work well); slides, standard 75 × 25 mm.; no. 2 thickness cover slips; slide labels; alcohol—95% ethyl is best, but methyl or denatured alcohol can be substituted; KOH solution, 10 g. KOH per 100 cc. of water; xylene; mountant (we prefer HSR: see Appendix), most prepared solutions are intended for histological work and are not viscous enough for whole mounts—it is best to prepare one's own from the crystalline form, using xylene.

1. PROCEDURE

Remove the abdomen and wet it well in alcohol. While the abdomen is soaking, fill a test tube with KOH and heat as described in the section on the head capsule of the cockroach. Again—*be very careful; KOH is an extremely caustic agent!* Soak the abdomen in the KOH solution for about an hour, then dissect as directed below. Alternatively, the abdomen may be left overnight in cold KOH solution, a procedure often making possible a more precise dissection because it does not completely destroy soft, unsclerotized structures.

MALE. Put the abdomen into a clean syracuse dish partially filled with tap water and add a few drops of alcohol. With a bent dissecting needle press shut the proximal end of the abdomen and with another press on the distal half of the abdomen, gradually working backward and forcing the contents out the rear of the abdomen. The heavily sclerotized structure exerted from the abdomen is the **genital capsule.** Alternatively, grasp a tergite in the forceps and use a fine scalpel, the blade turned outward, to slit the pleural conjunctiva along one side until the genital capsule is exposed. Before proceeding with the dissection refer to the figure of the male genitalia in Fox and Fox (1964: 108, Fig. 3.21) and become familiar with the terminology. Place the syracuse dish under the dissecting microscope and continue all dissections under magnification. Hook a needle under the **vinculum** and pull the genitalia free from the abdomen. With forceps clean away extraneous soft tissue, leaving only the sclerotized genital capsule. The best method of mounting the male genitalia of butterflies for study is to remove the left **valve** and the **penis,** then mount the remaining parts in profile with the other valve and the penis mounted separately on the slide. Note that the male genitalia of certain moths and other insects are sometimes mounted in posterior aspect by spreading and flattening the valves and uncus.

The three pieces of the genital capsule should now be placed in a syracuse dish of alcohol to remove water. The free valve and the penis may be allowed to float unobstructed in the alcohol, but it is desirable to weight down the other structures in the position of eventual mounting with a bit of glass. The parts should be left in the alcohol for fifteen to sixty minutes until dehydration is complete, then transferred to xylene—which substance is compatible with the mountant, whereas alcohol is not. Ten to twenty minutes in the xylene bath is generally sufficient.

Prepare a slide with a label giving all pertinent data, and then spread a thin layer of mountant in the center of the slide. Orient the pieces of the genitalia in this layer of mountant as desired, put three chips of glass around the pieces to serve as stilts for the cover slip and set aside until tacky. Add another drop of mountant to the area and place a cover slip over it. With care the preparation can be examined immediately, but it is generally best to let it dry a day or two.

FEMALE. The basic steps are as in the preparation of the male genitalia, but the dissection itself is entirely different. Carefully *dissect* away all the abdominal sclerites except the last two or three. Carefully pick away all the "detritus," but be careful not to remove reproductive organs and ducts (see Fox and Fox, 1964: 107, Fig. 3.20; 174, Fig. 5.9). Because the female genitalia are generally studied in ventral view, a glass weight to flatten the structures in the proper way is often necessary during the alcohol and xylene baths to ensure a well-oriented mount.

2. OBSERVATIONS

MALE. The dorsal portion of the male genitalia is composed of several elements, the most anterior of which is the **tegumen** (Tg, Fox and Fox, 1964: 108, Fig. 3.21) which connects to the **vinculum** (Vm) ventrally. Arising from the distal end of the tegumen and frequently fused to it is the **uncus** (Un), which may either be single (medial) or bifid (partly bilateral). From the ventro-lateral part of the tegumen and uncus arises the **gnathos** (plural, **gnathoi**), sometimes bilaterally paired. Follow the vinculum down to its ventral end; the structure projecting anteriorad into the abdomen is the **saccus** (Sac). Connected to the vinculum and projecting posteriad are the paired articulated **valves,** the structures that hold the female during copulation. The aperture in the anterior end of the penis is the **foramen,** through which the **ejaculatory duct** (Ejd) from the testes enters the penis; in some species spine-like **cornuti** will be seen either in the penis or on a membrane exserted from its posterior end. These structures are frequently of specific value for identification.

FEMALE. The structures most readily seen are the posterior paired lobes, the **papillae analae,** and the more anterior **vaginal plate.** In insects having a true ovipositor, these structures correspond to the processes from the eight and ninth abdominal segments shown by Fox and Fox (1964: 107, Fig. 3.20); the opening of the ovipositor is located between them. The duct leading internally from the ovipositor is the **egg exit duct** (Egd in Fox and Fox, 1964: 174, Fig. 5.9), which in turn leads into the **bursa copulatrix** (BCop) where the sperms are stored. This last structure is generally the innermost to survive KOH treatment and the ducts leading from the **ovaries** (O) and entering the bursa copulatrix, where fertilization of the oötids is accomplished, are usually dissolved in KOH.

F. MORPHOLOGICAL VARIATIONS

Almost no structure is constant in form throughout the Insecta. The generalized morphology of the cockroach serves well enough as a frame of reference with which to compare the almost infinite variations of the parts of insects. In principle, the specific structures of the roach may be taken as being rather near the hypothetical starting point. Time and available material must dictate the extent to which comparisons along these lines are made. It is not enough merely to notice morphological variation, however; one should constantly ask oneself *how* an observed modification benefits the animal possessing it.

The cockroach also may be taken as a fundamental insect type with which to compare the morphology of myriapods and of arachnids. In making such comparisons it is desirable for a student to list the points of similarity and the points of divergence between the classes.

MATERIALS. Dissecting microscope; compound microscope; dissecting forceps and needles; watch glasses; materials and media for making whole mounts, as desired (see Appendix).

Larger specimens are best preserved in 70% alcohol to which a little glycerin has been added, though pinned specimens may also be used. Small specimens, such as parasitic insects, mites, early instar larvae and the like can be prepared as whole mounts, using the technique detailed above for genitalia. Preparation of such mounts may also be used as a class project.

For study of the major variations in insect antennae, mouthparts, wings, legs, gills and cerci as outlined below, the following specimens may be provided: Ephemerida, adults; Odonata, adults and nymphs; Dictyoptera, adult mantids; Orthoptera, including Acrididae and Gryllotalpidae; Dermaptera; Anoplura, preferably the crab louse; Hemiptera, including Tingidae, Cynidae, Gerridae, Cicadidae; Coleoptera, adults, including Buprestidae, Scarabaeidae, Gyrinidae, Staphylinidae and Hydrophilidae; Neuroptera, *Corydalis* larvae; Trichoptera, adults; Lepidoptera, including adults of both sexes of a saturniid moth such as *Hyalophora cecropia* and of the monarch or other nymphaloid butterfly; Diptera, adult males of Culicidae or Chironomidae and larval Culicidae.

Dissected and mounted mouthparts and the hindlegs of a worker honeybee, *Apis mellifera;* mouthparts of a butterfly (Lepidoptera), of a flea (Siphonaptera), of a bug (Hemiptera), of a female mosquito (Diptera).

Preserved specimens of a scorpion (Scorpionida), spider (Araneida), tick or mite (Acarida), centipede (Chilopoda), milliped (Diplopoda).

1. INSECTS

HEAD. The commonest orientation of the mouthparts, **hypognathous** mouthparts, were noted in the cockroach and the mealworm. The larva of the dobson fly *Corydalis,* the "hellgrammite" of anglers, has the mouthparts diverted forward, rather than downward; this orientation of the mouthparts is called **prognathous.** Look at the head from the ventral aspect—the large, quadrangular flattened sclerite between the base of the labium and the cervix is the **gula,** characteristically well developed in insects with prognathous mouthparts. Now compare the head and mouthparts of a hemipterous insect. The mouthparts are diverted posteriad, but the basic structures of the head are modified chiefly only in proportion. Such mouthparts are known as **opisthognathous**, a form which is merely a modification of the hypognathous condition. A fuller discussion of the relationships between these mouthpart orientations is given by Fox and Fox (1964: 50).

The antennae exhibit numerous modifications throughout the insects, several of which we have seen. The antennae of the cockroach are **filiform,** those of the grain beetle **clavate** and those of the fruit fly **aristate.** These and other types of antennae are figured by Fox and Fox (1964: 78, Fig. 3.8), and the illustrations referred to in this paragraph are from that plate. Examine a larger buprestid beetle (Coleoptera: Buprestidae) with **serrate** antennae (Fig. 3.8E). All the Scarabaeoidea (Coleoptera) are characterized by **lamellate** antennae (Fig. 3.8H) as may readily be seen in one of the familiar June beetles, *Phyllophaga* species. A male mosquito or a midge (Diptera: Culicidae and Chironomidae, respectively) show yet another type of antenna, related to that of *Drosophila,* the **plumose** antenna. Examine a male saturniid moth, such as *Hyalophora cecropia* (Drury), and note the extreme type of plumose antenna known as **doubly bipectinate.** If a female moth is available, compare her antennae to those of the male—an easy method of determining sex. There are other modifications of the antennae, but the ones mentioned above will be the most frequently encountered.

Only two types of mouthparts have thus far been presented, the chewing mouthparts of *Blaberus* and *Tribolium* and the sponging mouthparts of *Drosophila*. The mouthparts of the honeybee, *Apis mellifera* (Linné), are quite remarkably modified, as shown in Fox and Fox (1964: 91, Fig. 3.14B). Examine a slide of these mouthparts and identify the parts indicated in the figure. The **proboscis** of the butterfly (Lepidoptera), shown by Fox and Fox (1964: 92, Fig. 3.15A), is composed of only the **galeae** of the maxillae. The maxillary palpi are absent, but the **labial palpi** are present; all other parts are missing in most Lepidoptera. Examine the mounted mouthparts of a flea (Siphonaptera) and compare the structures seen with those shown by Fox and Fox (1964: 96, Fig. 3.17). All the parts except the mandibles are involved in this proboscis. Compare these mouthparts with those of a female mosquito, shown by Fox and Fox (1964: 94, Fig. 3.16B), in which all the basic elements are present in the sucking proboscis. The feeding structures of a hemipterous insect also include all the basic parts, as shown by Fox and Fox (1964: 99, Fig. 3.18B), but the arrangement of the mouthparts of these insects is remarkably complex.

The whirligig beetles (Coleoptera: Gyrinidae) have remarkable eyes in that they are divided into dorsal and ventral halves, a striking adaptation to the habits of the gyrinid, which swims on the surface of the water and seeks its prey beneath it. The dorsal halves of the eyes are used above the water, the ventral halves below.

THORAX. Both membranous hindwings and leathery forewings have been noted in the cockroach. The forewings of *Tribolium* and other Coleoptera are hard and form a protective shield over the dorsal surface of the abdomen, a modification known as an **elytron.** Similar in function is the **hemelytron** of the heteropterous Hemiptera. Examine the hemelytron of several Heteroptera and note the variation in sclerotization from the highly reticulated hemelytron of the Tingidae (lace bugs) to the almost completely sclerotized wing of the Cynidae (Negro bugs).

Certain beetles have greatly shortened elytra, as in the rove beetles (Staphylinidae), some of which have as many as six abdominal tergites exposed. Membranous wings, however, are the general rule in most insects. In *Drosophila* the number of true veins and cross veins is very small. At the other extreme, members of the orders Odonata, Ephemeroptera and Neuroptera have a great many veins, and assigning homologies to them is extremely difficult. Another type of wing is found in the Lepidoptera and Trichoptera, which respectively have scales or hairs covering the membranous areas. Examine examples of each under the dissecting microscope. Remove some of the scales or hairs with a fine brush; place them in water on a slide under a cover slip, and examine them under the compound microscope. The configuration of the hairs and scales is characteristic of certain groups. Can you see why these two orders of insects are considered "close" to one another evolutionarily?

In the insects thus far studied in this chapter, the legs have been relatively unmodified. A number of specializations are shown by Fox and Fox (1964: 74–75, Figs. 3.6 and 3.7). A fairly generalized leg is shown by the grasshopper (Orthoptera) hindleg, the femur of which is greatly enlarged and has powerful muscles for the jumping locomotion from which the common name of this insect is derived (see Fig. 3.6B). The hindleg of a hydrophilid beetle (Fig. 3.6C) is, by contrast, highly modified: the entire leg is streamlined and the tarsus is provided with long, stiff hairs which collectively function as an oar to propel the beetle through the water. The midleg of the water strider (Hemiptera: Gerridae) is modified to suit its skipping motion across the water; note the great fringe of branched hairs arising from the tip of the first tarsal segment. Many modifications are found in the forelegs of insects: that of the mole cricket (Orthoptera: Gryllotalpidae, genus *Neocurtilla*) is very strong and is adapted for burrowing; the foreleg of the mantis is enlarged and provided with numerous spines for grasping prey; the foreleg of the crab louse (Anoplura) is equipped with hooks for gripping its host's hairs (see Fig. 3.6D, G, I). The foreleg of a nymphaloid butterfly, such as the monarch *Danaus plexippus* (Linné), is aborted and does not participate in locomotion, though it probably has other functions. A most remarkable leg modification is found in the honeybee, as shown in Figure 3.7A. The outer surface of the tibia is provided with rows of high, curved setae which form the "pollen basket" (**corbicula**). This structure is used for storing pollen on foraging expeditions and is present only in workers.

The wings and the legs serve together for stridulation in some of the Orthoptera, which make their sounds by rubbing the hindlegs against the wings (**stridulation**). The sounds so produced are highly specific, as will be seen in Chapter Two.

Abdomen. **Gills** are present in various aquatic larvae, often as segmental elements, but there are other arrangements. The **respiratory siphons** of mosquito larvae are unique, but the **anal gills** of damselflies (Odonata: Zygoptera) are perhaps the most spectacular of all insect gills. Examine as many types of gills as time and material permit; gills are useful in recognizing the various aquatic nymphs.

The **cerci** are modified in several ways. Cerci are normally formed in the cockroach but are absent in *Drosophila*. Examine an earwig (Dermaptera); the long forceps-like cerci are used as defensive weapons. In the mayfly (Ephemeroptera) the cerci are long, filamentous structures, frequently longer than the body; the similar-looking structure between them is the **median filament,** a prolongation of the eleventh tergite.

2. MYRIAPODA

Living myriapods, though highly specialized, provocatively suggest a fairly close relationship with insects. The body plan of myriapods (see Fox and Fox, 1964: 44–47) is generally said to consist of a head capsule and a long trunk made up of similar segments. Yet, examination of the anterior part of the myriapod trunk demonstrates that the first few segments differ to some degree from the rest of the trunk; this anterior region is a thorax in the making, as it were.

With their highly specialized mouthparts, their strongly thorax-like anterior trunk segments and the fusion of the rest of the trunk into diplosegments, the Diplopoda are more advanced, in these respects, than the Chilopoda. For a discussion of the legs of myriapods, see Fox and Fox (1964: 68–69). In most species the segments of the leg are easy enough to identify by counting out from the body and paying attention to the planes of articulation. Note the poison claws, modified legs of the first trunk segment, in the centipede and notice the smaller size of the legs on the anterior few trunk segments. Mouthparts are summarized by Fox and Fox (1964: 82–86). The feeding appendages of the milleped and the chilopod should be dissected and prepared as whole mounts for comparison (see Fox and Fox, 1964: 83–84, Figs. 3.11 and 3.12). Note the primitive, two-segmented mandibles in both kinds of myriapod and the leg-like maxillae of the chilopod. The **gnathochilarium** of the diplopod probably is the result of fusion of the first and second maxillae. In view of its highly specialized construction, it is easy to understand why it has been difficult to work out the true homologies of this part.

3. ARACHNIDA

There is no truly primitive form among the living Arachnida—perhaps the scorpion preserves more presumably primitive features than are found in other orders of the class. By comparison, both the spider and the tick are strongly modified. The loss of clear body segmentation in these two groups, as compared to the well-marked segmentation of the scorpion body, leads one to expect evolutional modification in other structures as well.

The arachnid body plan is discussed and figured by Fox and Fox (1964: 56–61); the arachnid appendages are considered by the same authors (1964: 64–67, 79–82, 109–111). These discussions should be consulted in comparing examples of the class. Note especially the various functional specializations of the five pairs of legs and the fundamental presence of two trochanters. Observe the variations in size and position of the eyes. Under the high power of the dissecting microscope and lower powers of the compound microscope, observe the various kinds of integumental setae, many of which are sensory.

CHAPTER TWO

Systematics

The first step toward experimentation with animals is to become familiar with the morphology of the animal to be investigated; the second step is to become familiar with the name of the experimental animal and its place in the animal kingdom relative to all other animals. Experimental work conducted without morphologic and taxonomic knowledge is as a house built upon the sand: no matter how large the supporting funds or how complex and expensive the equipment used, improperly grounded experiments are not experiments at all but only busywork.

Literature on the philosophy of systematics is very large. Primary modern references include Dobzhansky (1951), Mayr (1947) and Simpson (1961). Discussions on systematics and evolution are to be found in most books on evolution (Dodson, 1960; Ross, 1962; Ehrlich and Holm, 1963). A summary is presented for student information by Fox and Fox (1964: Chapter Nine).

However, most workers are concerned with the use of taxonomic results rather than with the philosophy of classification. The purpose of the present chapter is to acquaint the student with the use of taxonomic tools, of which the commonest and most valuable is the **key.** In the sections following, the construction and use of keys are discussed; practice keys at class/order, at family and at generic level are given, and the student is introduced to methods of identification employing behavioral and physiological characteristics.

A common misconception to the contrary notwithstanding, taxonomy is today a dynamic and exciting area of investigation. The number of new species and subspecies being described currently is as great as or greater than ever in the past. The impact of experimental biology, of the study of evolution and of new methods and viewpoints is leading to a profound reorganization of taxonomic concepts at all levels. One aspect of these developments is that keys to genera and species and even to higher categories and which fully reflect contemporary knowledge are hard to find and often are not yet in existence. Insect identification, in the modern scheme of things, has become less effective as a vehicle for introducing the beginner to entomology; it has become an advanced area of study.

It is not possible within the space available to provide keys enabling the student to identify any insect, myriapod or arachnid he happens to have: in fact, assembling such keys would be a life work and printing them would require a dozen or more volumes. Even a list of references to source keys for the identification of all terrestrial arthropods is well beyond the scope and intent of this book. However, the annotated references in the Appendix should be helpful for general use.

A. KEYS

A taxonomic key is a table or synopsis of characteristics of the members of a group, organized in such a way that it can be used to identify group members. In contemporary expertese a key is an "information retrieval system." There are various ways of constructing keys, and there are some pitfalls that may entrap the unwary writer of keys.

To afford a concrete example, consider the classes of terrestrial arthropods. The Insecta, Myriapoda and Arachnida are regarded as being terrestrial even though some of the life stages may be aquatic and some of the adults may live in the water: all but a few immatures utilize atmospheric oxygen. In contrast, the class Crustacea includes species that are characteristically aquatic and utilize oxygen dissolved in the water; exceptional among Crustacea are some members of the order Isopoda, the pill bugs and sow bugs. These Isopoda must be included in a consideration of terrestrial arthropods.

One way to present the characteristics of these groups is by use of a tabular outline form, thus:

A. Two pairs of antennae present. Class CRUSTACEA.
- B. With oval, dorso-ventrally flattened body; found in damp terrestrial habitats. Some species of the order ISOPODA.
- BB. Otherwise, not terrestrial. All other CRUSTACEA.

AA. One pair of antennae and a distinct head present.
- B. With 3 pairs of legs. Class INSECTA.
- BB.. With more than 3 pairs of legs. Class MYRIAPODA.

AAA. Antennae absent.
- B. Body divided into cephalothorax and abdomen, with no distinct head. Class ARACHNIDA.
- BB. Body divided into head, thorax and abdomen. Order PROTURA of the class INSECTA.

The tabular method of preparing a key works well enough, and the outline above would serve adequately for the identification of the classes of terrestrial arthropods. Note that the Protura, usually assigned to the class Insecta, are aberrant in that the antennae are lacking. Because the number and presence of antennae was used as the primary characteristic for distinguishing the classes—the three lines beginning with A—Protura had to be entered separately even though it is not a class. To use this kind of key, first locate the line beginning with an A which is true of the specimen being identified; then locate beneath the true A line the line beginning with B which is true for the specimen. For a short outline like the one given, this kind of key is easy to use. If there is a large number of categories in a key, the tabular method becomes exceedingly cumbersome and difficult to use.

For an example of the tabular key used for numerous categories, see Comstock (1940: 212–216). There are two A lines, one at the top of the first page of the key, the other (AA) in the middle of the second page; under the first A line the BB categories require all the letters to K, and under the second A line all the letters to S are used. Notice also that the need for indenting the successive entries from the left margin wastes space. An even more cumbersome example is to be found in Rothschild and Jordan (1898–1899) in a paper begun in one volume of a journal and continued in the next. An outline key to species begins in the first part of the study with an I category and its alternate II is in the next volume; on the next line under I is category A and its alternative AA is finally located thirty-four pages later.

The outline or tubular method is useful in one's own notebook as a preliminary step to preparing a more usable key. The couplet type of key has become usual because it is so much more easily followed. The data from the outline key to terrestrial arthropods can be converted to couplets quite easily:

1. Two pairs of antennae present; body oval, flattened dorso-ventrally; found in damp habitats .... (class CRUSTACEA) order ISOPODA
   One pair of antennae and a distinct head present 2
   Antennae absent 3
2. With 3 pairs of legs .... class INSECTA
   With more than 3 pairs of legs .... class MYRIAPODA
3. Body divided into cephalothorax and abdomen, no distinct head .... class ARACHNIDA
   Body divided into head, thorax and abdomen .... (class INSECTA) order PROTURA

To use a couplet key, enter the first set of alternatives and find which is true of the specimen being identified. Let us say that the specimen is an insect, so that the second of the three alternatives will be true; at the end of the line one is directed to couplet 2, and because the specimen has only three pairs of legs and no more, it is identified as class Insecta.

The above couplet key is a direct conversion from the outline key, but several refinements are possible. The first couplet comprises three alternatives; many people insist that all couplets be dichotomous, though sometimes it is convenient to use more than two alternatives. Also, a key is more easily used if the alternatives of each couplet are set off; one method of doing this is to designate the first alternative as A and the second as B. Another possible improvement is to make it easier to "run the key backward," that is, to start with an assumed known in the body of the key and work back by way of verification—to do so, the number of the couplet that leads to the couplet in question is entered in parentheses. With all three of these improvements incorporated into the key to terrestrial arthropod groups, the resultant version would be as follows:

1A. Two pairs of antennae present; body oval, flattened dorso-ventrally; found in damp habitats . . . . (class CRUSTACEA) order ISOPODA
1B. Fewer than 2 pairs of antennae present, or none 2
2A. (1B) One pair of antennae and a distinct head present 3
2B. (1B) Antennae absent 4
3A. (2A) With 3 pairs of legs . . . . class INSECTA
3B. (2A) With more than 3 pairs of legs . . . . class MYRIAPODA
4A. (2B) Body divided into cephalothorax and abdomen, no distinct head . . . . class ARACHNIDA
4B. (2B) Body divided into head, thorax and abdomen . . . . (class INSECTA) order PROTURA

All couplets are now dichotomous, the alternatives are designated A and B and the source couplet is indicated in parentheses. To use this key for verification of a supposed identification, let it be supposed that one has a spider, a member of the class Arachnida. First check the characteristics noted in couplet 4A. Since they are found to be true, go to couplet 2B as directed by the notation in parentheses and check the characteristics noted. Since these, too, are found to be true, go to couplet 1B, also found to be true. The tentative identification has been verified. Note, however, that it is possible to "run a key backward" even when the references are not included by finding the couplet numbers along the right-hand margin. Many keys are written without parenthetic reference numbers and often a key does not have the alternatives lettered.

The principal pitfall in writing a key is the use of such expressions as "large" alternating with "small," "longer" versus "shorter" and the like. The user of the key cannot be expected to know how large is "larger" or the length of "longer." The language in a key must be clear, unambiguous and specific.

1. KEY TO THE CLASSES AND ORDERS OF TERRESTRIAL ARTHROPODS

The following key at the class/order level is based in part on the excellent *A Key to the Principal Orders and Families of Insects* by Z. P. Metcalf and C. L. Metcalf. By permission of the copyright owners, the material pertaining to insect orders has been used with minor modifications, and to this we have added original material pertaining to the myriapods and arachnids. The Metcalf and Metcalf *Key* is well illustrated, very easy to use and is one of the finest of its kind to come to our attention; the student of entomology will find it invaluable.

KEY TO THE CLASSES AND ORDERS OF TERRESTRIAL ARTHROPODS (ADULTS ONLY)

1a. Two pairs of antennae present; body oval, flattened dorso-ventrally; found in damp habitats . . . . class CRUSTACEA, order ISOPODA
1b. One pair of antennae present or antennae absent 2
2a. Antennae absent 3
2b. One pair of antennae and a distinct head present 13
3a. No distinct head; body divided into cephalothorax and abdomen . . . . (class ARACHNIDA) . . . 4
3b. Body divided into head, thorax and abdomen; eyes lacking . . . . class INSECTA, order PROTURA
4a. First leg (pedipalpus) bearing an enlarged claw 5
4b. First leg bearing a tiny claw or none 6
5a. Pectines present; abdomen with anterior segments about the same diameter as cephalothorax, but 5 posterior segments smaller and tail-like, tipped with a recurved telson containing poison glands . . . . order SCORPIONIDA
5b. Pectines absent; abdomen oval, not divided into 2 distinct sections . . . . order PSEUDOSCORPIONIDA
6a. Body more or less oval, the anterior abdominal segment or segments not constricted, abdomen not well-separated from cephalothorax 7
6b. Anterior abdominal segment or segments strongly constricted, the abdomen and cephalothorax distinctly separated 9
7a. Anterior cephalothoracic segments forming a false head (gnathosoma), bearing the chelicerae and pedipalpi . . . . order ACARIDA
7b. Anterior cephalothoracic segments not forming a gnathosoma 8
8a. Legs proportionately very long and slender; always at least several times the length of the body . . . . order PHALANGIDA
8b. Legs proportionately short and stout, none longer than the body . . . . order RICINULEIDA
9a. Abdomen tipped with a tail-like, often multiarticulate telson 10

9b. Tip of abdomen rounded; telson, if evident, not tail-like 11
10a. One-half to 3 mm. long; first legs (pedipalpi) similar to other legs, slender; chelicerae pincer-like with 3 segments .... order PALPIGRADIDA
10b. Two to 65 mm. long; first legs much stouter than other legs; chelicerae with 2 evident segments, the "fixed finger" reduced .... order THELYPHONIDA
11a. Second pair of legs very slender, greatly elongated and antenna-like; body flattened dorso-ventrally; first legs powerful, tipped with a spine or hook .... order PHRYNEIDA
11b. Second pair of legs normal, not antenna-like; body not flattened dorso-ventrally 12
12a. First pair of legs (pedipalpi) short, sensory, not used in locomotion; spinnerets present .... order ARANEIDA
12b. First pair of legs as long as or longer than the fifth; spinnerets not present .... order SOLPUGIDA
13a. More than 3 pairs of legs present .... (class MYRIAPODA) ... 14
13b. Three pairs of legs present .... (class INSECTA) ... 17
14a. Never more than 1 pair of legs on a segment, the trunk segments not fused in pairs 15
14b. Many apparent trunk segments bearing 2 pairs of legs, most trunk segments fused in pairs (diplosegments) 16
15a. First pair of legs stout and enlarged (poison claws), carried beneath the head; reproductive system opening at or near the posterior end of the trunk .... subclass CHILOPODA
15b. First pair of legs not enlarged or modified as poison claws; reproductive system opening on third evident trunk segment .. subclass SYMPHYLA
16a. No more than 1.5 mm. long; antennae apparently branched .... subclass PAUROPODA
16b. More than 2 mm. long; antennae not branched .... subclass DIPLOPODA
17a. Wings not present 18
17b. Wings present 33
18a. Abdomen with 6 or fewer apparent segments, provided with a ventral furcula .... order COLLEMBOLA
18b. Abdomen with more than 6 apparent segments and without furcula .. 19
19a. Styli present on some of the abdominal sternites 20
19b. Styli never present on abdominal sternites 21
20a. Tip of abdomen bearing 3 long multisegmented processes, the median caudal filament and 2 cerci .... order THYSANURA
20b. Median caudal filament absent, the cerci either multisegmented or modified as forceps .... order APTERA
21a. Mouthparts fitted for chewing 22
21b. Mouthparts fitted for piercing, lapping or sucking and sometimes concealed 27
22a. Louse-like insects 23
22b. Insects of various forms, not louse-like 24

23a. Antenna with 5 or fewer segments . . . . order MALLOPHAGA
23b. Antennae with more than 5 segments . . . order CORRODENTIA (part)
24a. Abdomen with cerci 25
24b. Abdomen without cerci 26
25a. Cerci developed as heavily sclerotized forceps . . . order DERMAPTERA
25b. Cerci not developed as forceps, sometimes short 44
26a. Anterior part of abdomen constricted (pedicel) and much narrower than thorax . . . order HYMENOPTERA, families Formicidae and Mutillidae
26b. Anterior part of abdomen nearly as broad as thorax . . . . order COLEOPTERA, family Lampyridae (part)
27a. Small, scale-like insects, usually without legs, eyes or antennae and usually covered by a scale beneath which the insect lives. Sometimes legs and antennae are present, the scale wanting, and the insect covered with a mealy powder . . . order HEMIPTERA, family Coccidae (part)
27b. Insect not scale-like or sedentary; legs and antennae always present . . 28
28a. Tarsi with 5 segments 29
28b. Tarsi with fewer than 5 segments 31
29a. Body laterally compressed . . . . order SIPHONAPTERA
29b. Body not compressed, wider than high 30
30a. Abdomen indistinctly segmented, covered with hairs . . . . order DIPTERA, family Hippoboscidae (part)
30b. Abdomen distinctly segmented, covered with scales . . . . order LEPIDOPTERA, ♀'s of various families
31a. With 1 or 2 post-tarsal claws; mouthparts enclosed in a beak or withdrawn into the head 32
31b. Post-tarsal claws lacking; mouthparts not enclosed in a beak . . . . order THYSANOPTERA, family Thripidae
32a. Beak evident, segmented . . . . order HEMIPTERA (part)
32b. Beak not evident, mouthparts withdrawn into head . . order ANOPLURA
33a. With 2 wings 34
33b. With 4 wings 38
34a. Wings membranous 35
34b. Wings horny or leathery, not membranous 36
35a. Wings with few or no cross veins; halteres present . . . . order DIPTERA
35b. Wings with numerous cross veins . . . . order EPHEMERIDA (part)
36a. Mouthparts fitted for chewing 37
36b. Mouthparts fitted for piercing, enclosed in a beak . . order HEMIPTERA
37a. Wings leathery with many veins (tegmina); hindlegs fitted for jumping . . . . order ORTHOPTERA (part)
37b. Wings horny without veins (elytra) . . . . order COLEOPTERA, family Meloidae
38a. The two pairs of wings unlike in structure, *i.e.*, with forewings usually much thicker or less transparent than hindwings, or the forewings only slightly thicker and the hindwings folded like a fan when at rest . . . 39
38b. The 2 pairs of wings similar in structure, *i.e.*, of the same degree of thickness or transparency, and the hindwings not folded like a fan when at rest 51

39a. Forewings thick at base, thinner and overlapping at tip; mouthparts fitted for piercing, the beak arising from the front part of the head . . . . order HEMIPTERA, suborder Heteroptera
39b. Forewings of same texture throughout 40
40a. Forewings horny, without veins 41
40b. Forewings leathery or opaque, with veins 43
41a. Abdomen provided with forceps (modified cerci) at posterior end . . . . order DERMAPTERA (part)
41b. Abdomen without forceps 42
42a. Forewings meet in a straight line along the back when at rest . . . . order COLEOPTERA
42b. Forewings much shorter than hindwings, held horizontally from the body when at rest . . . . order STREPSIPTERA
43a. Mouthparts fitted for piercing, the beak arising from the ventro-posterior part of the head near the front legs . . . . order HEMIPTERA, suborder Homoptera
43b. Mouthparts fitted for chewing 44
44a. Tarsi with 2 segments, legs short . . . . order ZORAPTERA
44b. Tarsi with more than 2 segments 45
45a. First foretarsal segment enlarged, suboval, containing silk glands . . . . order EMBIOPTERA
45b. First foretarsal segment not thickened or enlarged 46
46a. Hindlegs longest, fitted for jumping; hindfemora large and strong . . . . order ORTHOPTERA
46b. Hindlegs no longer than midlegs; hindfemora not enlarged 47
47a. Antennae moniliform, legs short and similar, body ant-like . . . . order ISOPTERA
47b. Antennae filiform, body not ant-like 48
48a. Prothorax longer than meso- and metathorax together; forelegs with femora and tibiae provided with spines and teeth along their opposing edges . . . . order DICTYOPTERA, suborder Mantodea
48b. Prothorax not longer than meso- plus metathorax; all legs rather similar 49
49a. Body oval, flattened . . . . order DICTYOPTERA, suborder Blattaria
49b. Body not oval 50
50a. Body slender and stick-like or wide and leaf-like . . . . . order PHASMIDA
50b. Body cylindrical . . . . order GRYLLOBLATTODEA
51a. Post-tarsal claws present 52
51b. Post-tarsal claws absent . . . . order THYSANOPTERA
52a. Wings covered, at least partly, with scales . . . . . . order LEPIDOPTERA
52b. Wings not scaled 53
53a. Mouthparts fitted for piercing, formed as a beak . . . order HEMIPTERA
53b. Mouthparts not fitted for piercing, usually for chewing, sometimes for sucking or lapping, sometimes greatly reduced 54
54a. Wings with many cross veins and usually many longitudinal veins . . . 55
54b. Wings with few cross veins and usually not many longitudinal veins . . 60

55a. Abdomen with a pair of long caudal filaments .... order EPHEMERIDA
55b. Abdomen without long caudal filaments 56
56a. Antennae short and inconspicuous, awl-shaped ...... order ODONATA
56b. Antennae conspicuous, usually long, never awl-shaped 57
57a. Cerci present 58
57b. Cerci absent 59
58a. Tarsi with 2 or 3 segments; hindwings as large as or larger than forewings; antennae filiform .... order PLECOPTERA
58b. Tarsi with 4 or 5 segments; fore- and hindwings of equal size; antennae moniliform .... order ISOPTERA
59a. Head prolonged into a beak with chewing mouthparts at its tip; antennae filiform .... order MECOPTERA
59b. Head not prolonged into a beak; antennae various but usually filiform .. .. order NEUROPTERA
60a. Tarsi with 1, 2 or 3 segments 61
60b. Tarsi with 4 or 5 segments.. 62
61a. Hindwings folded like a fan; tarsi with 3 segments .... order ORTHOPTERA (some Gryllidae)
61b. Hindwings not folded like a fan; tarsi with 1 or 2 segments .... order CORRODENTIA
62a. Forewings longer and larger than hindwings ... order HYMENOPTERA
62b. Hindwings as large as or larger than forewings .. order TRICHOPTERA

2. KEY TO THE GENERA OF ROMALEINAE OF AMERICA NORTH OF MEXICO

In the classification of Orthoptera presented by Rehn and Grant (1961), the suborder Caelifera included the pygmy mole crickets (Tridactyloidea) and the short-horned grasshoppers (Acridoidea). In this system the superfamily Acridoidea is divided into four families, one of which is Acrididae. Acrididae includes "all of the economically important migratory and devastating locusts or grasshoppers, as well as thousands of species of more sedentary habits, these generally less destructive agriculturally" (Rehn and Grant, 1961). The Romaleinae, one of the three subfamilies of Acrididae, comprises the lubber grasshoppers and near relatives, grouped into forty-six genera, all confined to the Americas. Seven of these genera are found in America in the region north of Mexico and a key for their identification is given by Rehn and Grant (1961: 175–178), a complex and highly technical key prepared for specialists. An unpublished version of the same key, written by Grant and quoted by permission, appears below. Comparison of the two versions shows that the one below is greatly simplified. Although written for the nonspecialist, it contains a few technical words which may not be familiar to the student entomologist; these words are defined as follows:

Angulate (*adj.*), forming an angle, as when two margins or lines meet in an angle.

Arcuate (*adj.*), arched or bowed.

Carina, -ae (*noun*), an elevated ridge, not necessarily high or sharp; literally, a keel.

Fastigium, -a (*noun*), that portion of the dorsal surface of the head cephalad of the interocular space (Rehn and Grant, 1961); the extreme point or front of the vertex.

Pronotal disk, the flattened dorsal surface of the pronotum.

Triangulate (*adj.*), of a triangular shape.

Truncate (*adj.*), cut off squarely at the tip.

Five of the preceding words are in common use in descriptive entomology, one (pronotal disk) is less commonly used and one (fastigium) is used mainly by specialists on Orthoptera. Keys generally contain a technical vocabulary which must be understood accurately. The student should possess an adequate glossary, dictionary or encyclopedia of biological terms (the best for entomology is Torre-Bueno, 1937, unfortunately out of print); few professional entomologists even pretend to be completely familiar with every term used in the field.

1a. Fastigium, in dorsal view, triangulately produced 2
1b. Fastigium, in dorsal view, low arcuate or with oblique frontal carinae, but never triangulately produced 5
2a. Posterior border of pronotal disk truncate or very low arcuate. (Found in many areas of western United States) . . . . *Brachystola* Scudder
2b. Posterior border of pronotal disk angulate 3
3a. Vertex with a median carina, especially pronounced posteriorly near the occiput (California) . . . . *Drachotettix* Bruner
3b. Vertex with median carina lacking or obsolete 4
4a. When tegmina and posterior femora are in normal position, the apices of the former extend beyond three-quarters the length of the latter (southwestern United States) . . . . *Taenipoda* Stål
4b. When tegmina and posterior femora are in normal position, the apices of the former never reach three-quarters the length of the latter (southeastern United States to central Texas) . . . . *Romalea* Serville
5a. With an apical spine on the external margin of the posterior tibia 6
5b. Without an apical spine on the external margin of the posterior tibia (Imperial and Coachella Valleys, California) . . . . . . *Spaniacris* Hebard
6a. Apices of tegmina extending beyond the apices of the posterior femora when both are in normal position (arid areas of Arizona, Utah, Nevada and California) . . . . *Tytthotyle* Scudder
6b. Apices of tegmina not reaching the apices of the posterior femora when both are in normal position (parts of southwestern United States) . . . . *Phrynotettix* Glover

B. BEHAVIORAL AND PHYSIOLOGICAL TAXONOMY

In recent years, behavioral and physiological characteristics have become increasingly more important in systematics, have begun to appear in taxonomic keys and often are at least included with general ecological considerations in descriptive material.

As Mayr (1958) has pointed out, there is a close correlation between behavior and systematics. Ross (1962) has reviewed a number of examples and concluded that the only known cases of true biological differences, other than structure, between closely related species involve trophic specificity and behavior. Some behavior patterns are conservative and thus are quite valuable in taxonomy and in phylogenetic reconstructions.

Baldwin (1948) and others have called attention to biochemical differences among broad groups of organisms. Paper chromatography and electrophoresis have shown particular promise as tools for separating difficult species complexes (Buzzati-Traverso, 1952, 1960; Micks, 1954; Lewallen, 1957; Sibly, 1960), although intraspecific variation remains a significant problem.

1. BEHAVIORAL IDENTIFICATION

Behavioral traits of value in separating taxa have primarily involved patterns associated with mating. Such behavior patterns are particularly significant because species are most frequently defined—although very seldom demonstrated—in terms of reproductive isolation.

a. Caddis Fly Cases

Case-building behavior in the order Trichoptera has been used extensively for the separation of major subgroups (Vorhies, 1909; Lloyd, 1921; Wessenberg-Lund, 1943) and families (Ross, 1944; Pennak, 1952). Wiggins (1960) has separated generic groups of phryganeid larvae on the basis of two basic types of case construction and Hanna (1960) has recently shown that several basic patterns of case repair exist within groups of caddis fly larvae.

Based only on the type of larval case, it is possible to identify the nineteen families of Trichoptera known to occur in North America. An initial attempt was made by Betten (1934), who designated ten case types. After studying case building in the Trichoptera, it becomes apparent that evolutionary processes in different habitats have obscured many phylogenetic relationships at the family or even generic level. The families Limnephilidae and Leptoceridae are particularly diverse with regard to types of case construction and cannot be separated as unit groups from the remaining caddis fly families by case type. In fact, in many instances the type of case built by several species of a genus varies so widely that the generic case is difficult to define.

The exercise below affords comparison of a taxonomic separation based entirely on the behavioral traits of case building by Trichoptera, without the usual morphological separation.

MATERIALS. Various trichopteran larvae and cases. It is important that caddis fly larvae be collected from a wide range of fresh-water habitats—for collection methods, see Appendix. Care must be taken to keep larvae and cases correctly associated and to take detailed field collection notes, particularly regarding noncase and nonportable-case forms. Taxonomic keys to the families of North American Trichoptera based on morphological characters (Ross, 1944; Pennak, 1952; Ross, 1959; Denning, 1956). Dissecting microscope and various dissecting instruments.

METHODS. Identify the collection of larval Trichoptera using the key to case types given below.

After the separation has been made on the basis of case construction, sort the larvae to families using one of the morphological keys cited above and compare the results of the two separations.

Add drawings and notes for larvae and cases not represented in the key to case type.

A PARTIAL KEY TO THE FAMILIES OF NORTH AMERICAN TRICHOPTERA, BASED ON LARVAL CASE TYPE (MATURE LARVAE)

1a. Free ranging (predaceous) larvae, without case or net . . . . (Fig. 2.1, *Rhyacophila*) RHYACOPHILIDAE

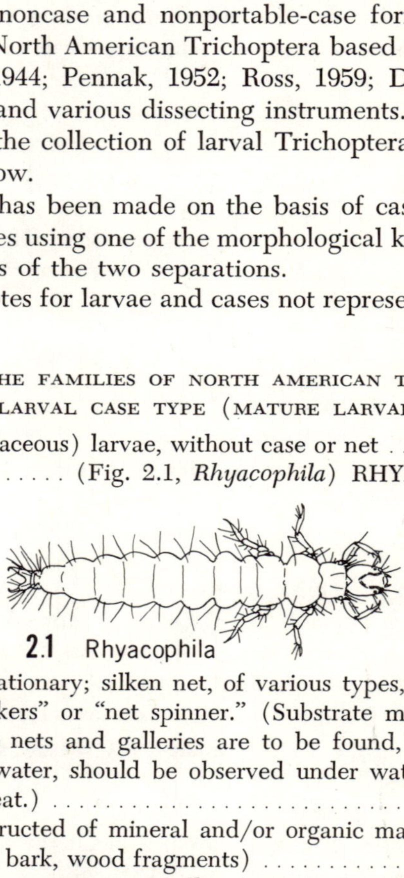

2.1 Rhyacophila

1b. Case, if present, stationary; silken net, of various types, always present. The "retreat makers" or "net spinner." (Substrate materials upon or under which the nets and galleries are to be found, *e.g.*, cobbles in rapidly flowing water, should be observed under water to avoid collapse of the retreat.) 2

1c. Portable case constructed of mineral and/or organic material (*i.e.*, leaf, stem, twig, stick, bark, wood fragments) 4

2a. (1) Case, if present, a stationary gallery or tube, loosely constructed of sand or fragments of leaf or wood and connecting with a broad semicircular, expanded net spun in the current (direction indicated by arrow) . . . . (Fig. 2.2, *Hydropsyche*) Hydropsychidae

2b. Net not a broad semicircular expansion, or no stationary case present . . 3

3a. (2) Nets low, long, flat, half-conical tunnels . . . . (Fig. 2.3, *Chimarra*) Philopotamidae

2.2 Hydropsyche

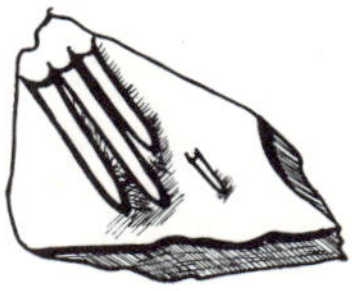

2.3 Chimarrha

3b. Nets conical in shape, curved at posterior end or with a bulbous posterior portion or constricted in the middle and expanded at both ends. If net is a semicircular expansion, then no stationary case present. One genus makes branched retreats buried, except for a small projecting turret, in the stream substrate . . . . (Fig. 2.4, *Polycentropus, Phylocentropus, Neuroclipsis, Wormaldia*) Psychomyiidae

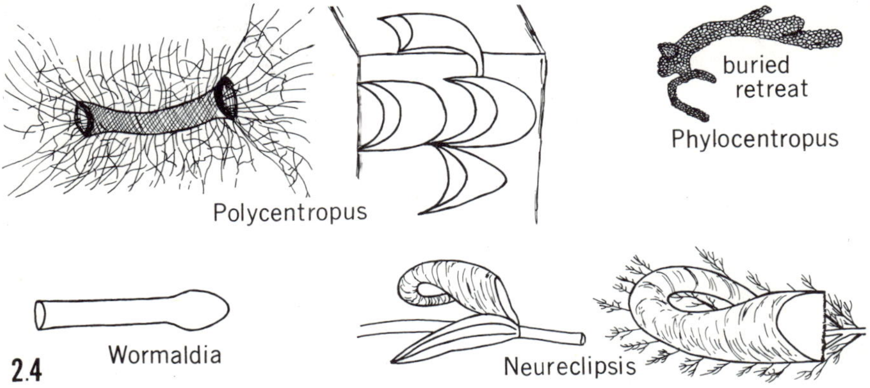

2.4

4a. (1) Case very minute (3–5 mm. in length), barrel- or purse-shaped, constructed of minute sand grains, fragments of organic debris or largely of silk. The "microcaddis flies" . . . . (Fig. 2.5) Hydroptilidae

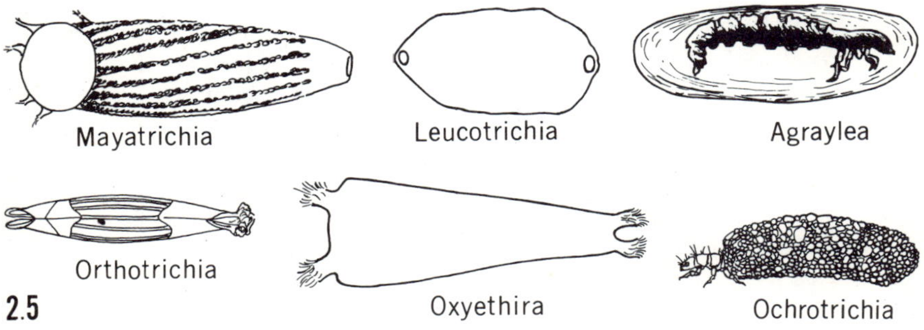

2.5

4b. Case of mature larva longer than 5 mm., not barrel- or purse-shaped . . 5

5a. (4) Case round or oval in dorsal view, without flanges 6

5b. General shape of case a cylindrical tube (may be bent or tapered, or square in cross section). If case is oval in dorsal view, shape due to lateral flanges 7

6a. (5) Case shaped like a snail shell, constructed of fine sand grains . . . . (Fig. 2.6, *Helicopsyche*) Helicopsychidae

6b. Case shaped like a tortoise shell (with a ventral strap), constructed of coarse sand grains . . . . (Fig. 2.7, *Glossosoma, Protoptila*) Glossosomatidae

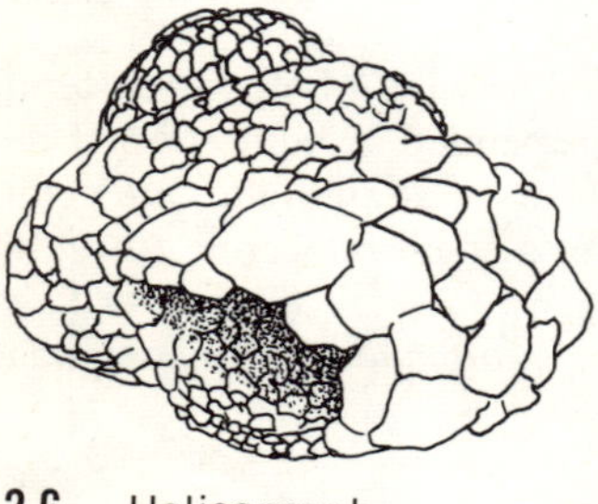

2.6 Helicopsyche

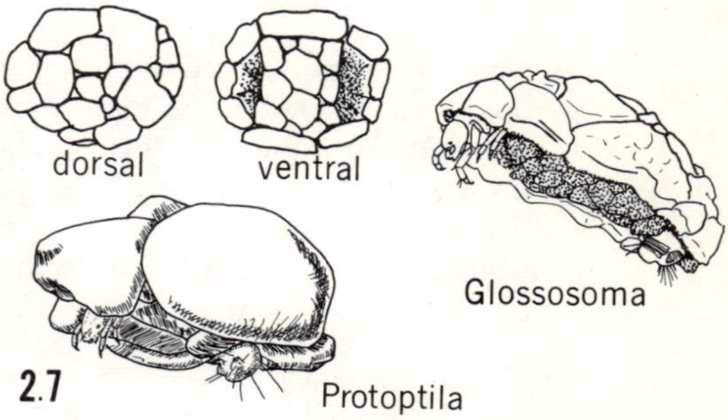

2.7

7a. (5) Case square in cross section 8

7b. Case round or triangular in cross section 9

8a. (7) Case square in cross section, markedly tapered from front to back; each square composed of 4 equal-sized fragments of stem or leaf material. (Some individuals in a population of a given species may have cases round in cross section over part or all of their length and composed largely of silk.) . . . . (Fig. 2.8, *Brachycentrus*) Brachycentridae

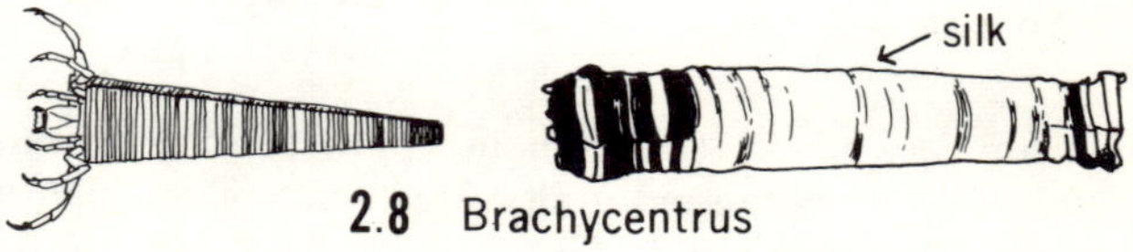

2.8 Brachycentrus

8b. Case irregularly square in cross section, tapered very little or not at all and composed of unequally sized vegetal fragments. (In populations of a given species there is a tendency for at least some individuals to construct cases of sand grains, round in cross section.) . . . . (Fig. 2.9, *Lepidostoma*) Lepidostomatidae (in part)

9a. (7) Case constructed entirely of organic material (*i.e.*, leaf, stem, twig, stick, bark, wood fragments) 10

9b. Case constructed with at least some mineral particles (may be entirely of mineral material or equally mixed with pieces of vegetation) 18

10a. (9) Construction a regular pattern; either a series of rings placed end to end or a spirally wound pattern; pieces of vegetation of fairly uniform size and shape 11

2.9 Lepidostoma

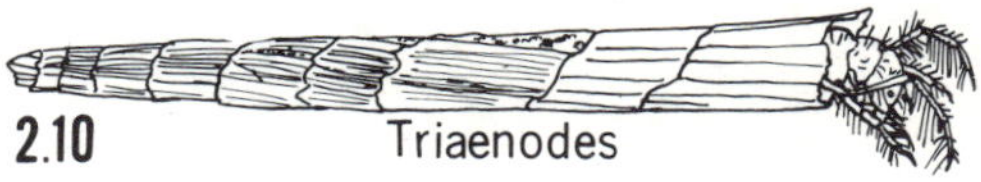

2.10 Triaenodes

10b. Case not of a regular ring or spiral pattern; pieces of vegetation of irregular size and shape. (Case may be of the "log-cabin" type or a hollowed-out twig or piece of wood) 12

11a. (10) Small cases, generally less than 15 mm. in length, with a spirally wound pattern of construction; conspicuous taper from front to back . . . . (Fig. 2.10, *Triaenodes*) Leptoceridae (1 genus)

11b. Large cases, generally 20 to 30 mm. in length, with either a ring (rings of *Ptilosomis* shown separated) or spirally wound (anterior end of *Phryganea* shown unwound) pattern; at most, only a slight taper from front to back . . . . (Fig. 2.11, *Ptilostomis, Phryganea, Agrypnia.*) Phryganeidae

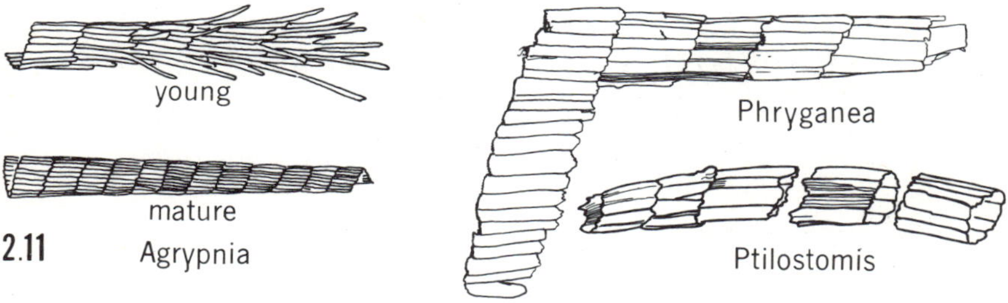

2.11 Agrypnia

12a. (10) Case a hollowed-out twig or piece of wood, open at both ends (15–25 mm. in length) . . . . (Fig. 2.12, *Heteroplectron*) Calamoceratidae (1 genus)

2.12 Heteroplectron

12b. Case not a hollowed-out twig or piece of wood; variously constructed of bits of vegetation 13

13a. (2) Case triangular in cross section 14

13b. Case round in cross section 15

14a. (13) Case usually composed of 3 large leaf pieces with some additional small vegetal fragments . . . . Calamoceratidae (part)

14b. Case usually composed of more than 3 smaller leaf or bark fragments. (Some members of a population may construct cases round in cross

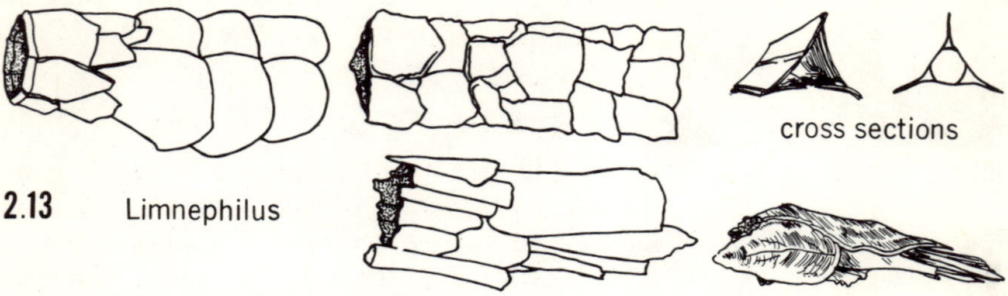

2.13 Limnephilus

section and occasionally of flat chips of mineral material.) . . . . (Fig. 2.13, *Limnephilus*) Limnephilidae (1 genus, in part)

15a. (13) Case "cross-stick" or rough "log-cabin" type of construction 16

15b. Case not log-cabin type 17

16a. (15) Fairly regular, small, log-cabin type case constructed of coarse stem (stick) material . . . . (Fig. 2.14, *Oecetis*) Leptoceridae (1 genus, in part)

16b. Quite irregular, large, log-cabin type case constructed of fine stem (grass) material . . . Limnephilidae (Platycentropus), or coarse stem (stick, seeds) material . . . (Fig. 2.15, *Platycentropus*) (*Limnephilus,* in part)

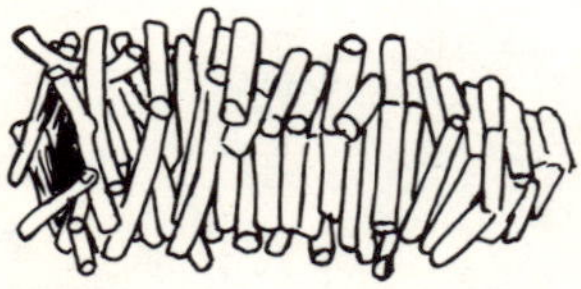

2.14 Oecetis

2.15 Platycentropus

17a. (15) Case with "balance-sticks" (usually two) along sides (cases of some young individuals may be broom-like clumps of sticks or stem pieces) . . . . (Fig. 2.16, *Pycnopsyche, Astenophylax*) Limnephilidae (in part)

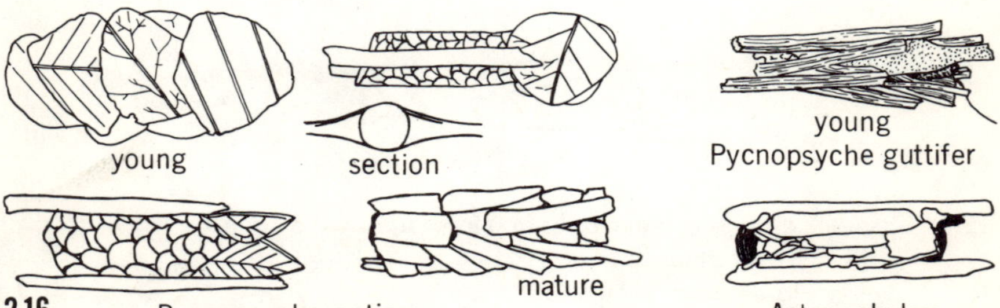

2.16 Pycnopsyche antica Astenophylax

17b. Case without "balance-sticks" (at most with fine, longitudinally placed fragments of stem or root), constructed of leaf and bark fragments . . . . (Fig. 2.17, *Glyphotaelius, Limnephilus*) Limnephilidae (in part)

18a. (9) Case constructed entirely of mineral material 19

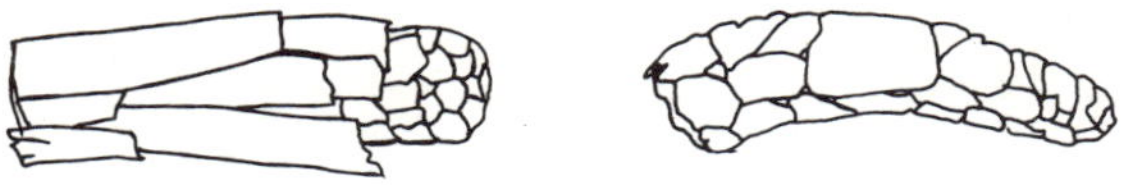

Glyphotaelius

2.17 Limnephilus sp. Limnephilus submonilifer

18b. Case of vegetal material (leaf, stick, stem, or bark fragments) with conspicuous amounts of mineral particles 29

19a. (18) Case with lateral flanges 20

19b. Case without lateral flanges 23

20a. (19) Case with dorsal projecting porch-like extension; flanges regular, may or may not be composed of particles more coarse than main tube of case 21

20b. Case without dorsally projecting "roof"; flanges irregular, composed of one or more coarse "ballast" particles on each side; case strong, slightly curved and flattened ventrally 22

21a. Case with dorsal projecting roof or flange anterior only and with lateral flanges of particles equal to or greater than the size of grains used in main tube . . . . (Fig. 2.18, *Molanna*) Molannidae

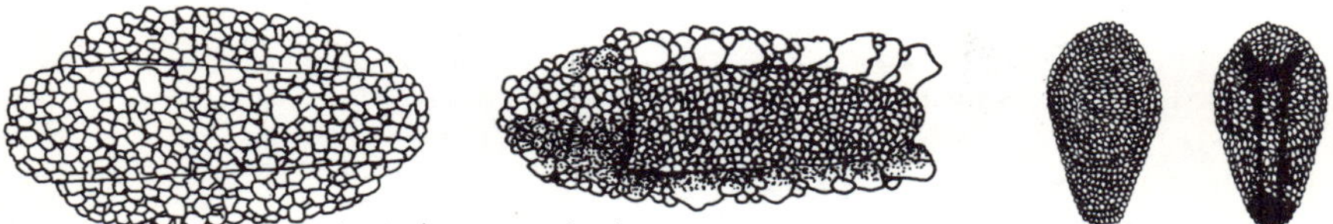

2.18 Molanna dorsal ventral dorsal ventral

21b. Case with flat dorsal surface projecting beyond posterior end of tube, as well as anteriorly and laterally. (These cases are difficult to separate from certain species of *Molanna*, especially those found in small springs.) . . . (Fig. 2.19, *Athripsodes*) Leptoceridae (1 genus, in part)

22a. (20) Case slightly larger (about 12 mm.), usually with two ballast particles on each side forming flanges equal to or greater than case length . . . . (Fig. 2.20, *Goera*) Goeridae

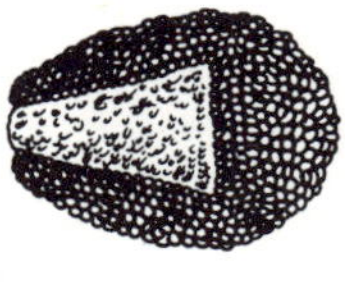

2.19 Athripsodes

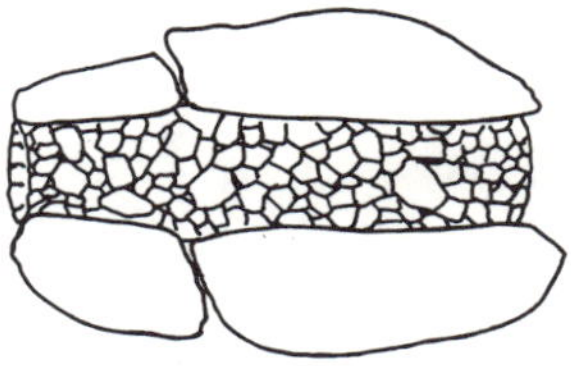

2.20 Goera

22b. Case slightly smaller (about 10 mm.), usually with several ballast particles on each side forming flanges shorter than case length .... (Fig. 2.21, *Neophylax*) Limnephilidae (1 genus)

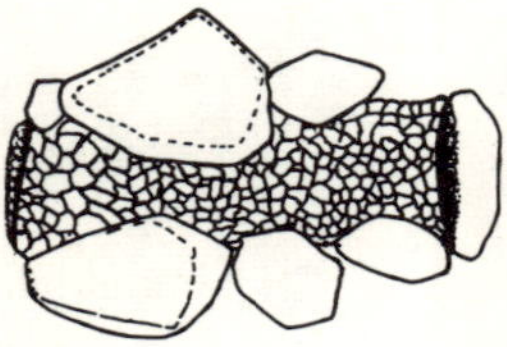

2.21 Neophylax

23a. (19) Case straight, not tapered (or only very slightly), constructed of fairly uniform-sized sand grains (all coarse or all fine) 24
23b. Case either curved or strongly tapered (cone-shaped) or both 25
24a. (23) Case large (generally greater than 15 mm. in length) and robust, usually constructed predominantly of particles 1 mm. or more in diameter, or of mollusk shell fragments ....
(Fig. 2.22, *Pycnopsyche lepida, Limnephilus*) Limnephilidae (in part)

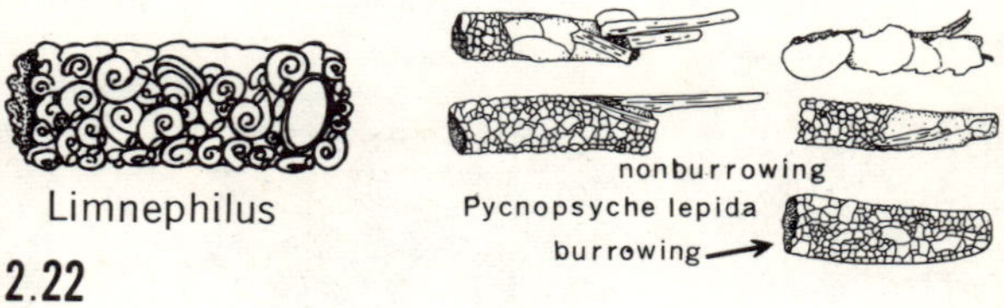

2.22

24b. Case strong, cylindrical, of fine sand (conical in early instars) .... Sericostomatidae
24c. Case small (generally less than 15 mm. in length) and delicate, usually predominantly of particles smaller than 1 mm. in diameter. (Some cases may be square in cross section; usually at least a few members of the population of a given species will have cases constructed of vegetal fragments and irregularly square in cross section—see 8b above) .... (Fig. 2.23, *Lepidostoma*) Lepidostomatidae (in part)
25a. (23) Case curved 26

2.23 Lepidostoma

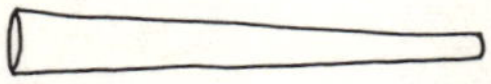

2.24 Leptocerus

Setodes

25b. Case straight and tapered (conical), long and delicate (case may actually be constructed primarily of silk in some species) . . . . (Fig. 2.24, *Leptocerus, Setodes*) Leptoceridae (in part)

26a. (25) Case tapered (conical) 27

26b. Case not tapered, strong, constructed of uniformly sized sand grains . . . . (Fig. 2.25, *Psilotreta*) Odontoceridae

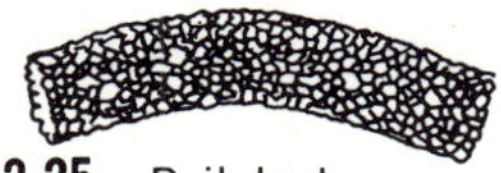

2.25 Psilotreta

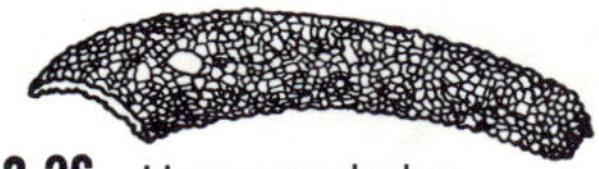

2.26 Hesperophylax

27a. (26) Case curved and conical with a somewhat pointed, hood-like dorsal projection at the anterior end which normally conceals the larva's head . . . . (Fig. 2.26, *Hesperophylax*) Limnephilidae (in part)

27b. Case without hood-like projection (or reduced, at least) 28

28a. (27) Case a long, delicate, curved, gradually tapering cone . . . . (Fig. 2.27, *Leptocella*) Leptoceridae (in part)

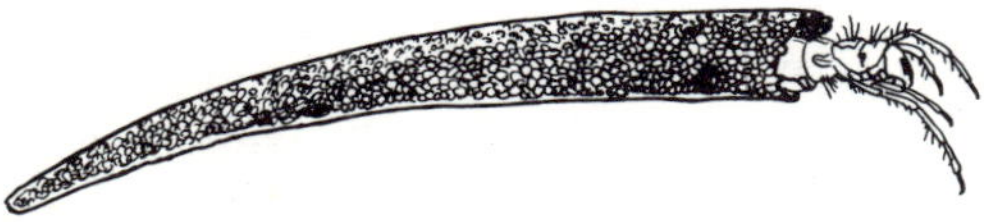

2.27 Leptocella

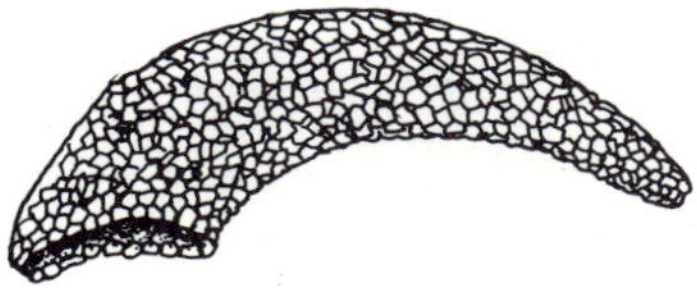

2.28 Athripsodes

28b. Case a short, stout, rather blunt, abruptly tapering cone (about 10 mm. long) . . . . (Fig. 2.28, *Athripsodes*) Leptoceridae (in part)

28c. Case conical, curved, gradually tapering, made of fine sand (about 6 mm. long) . . . . (Fig. 2.29, *Beraea*) Beraeidae

29a. (18) Balance sticks extending beyond anterior and posterior margin of

2.29 Beraea

2.30 Mystacides

case. (Some members of the population may have cases largely of organic material, but the normal mode of construction involves significant amounts of mineral material in case tube.) . . . . (Fig. 2.30, *Mystacides*) Leptoceridae (1 genus)

□

29b. Balance sticks extending beyond posterior margin only 30

30a. (29) A few large mineral particles used in construction of anterior one-third of case . . . . (Fig. 2.31, *Pycnopsyche guttifer*) Limnephilidae (in part)

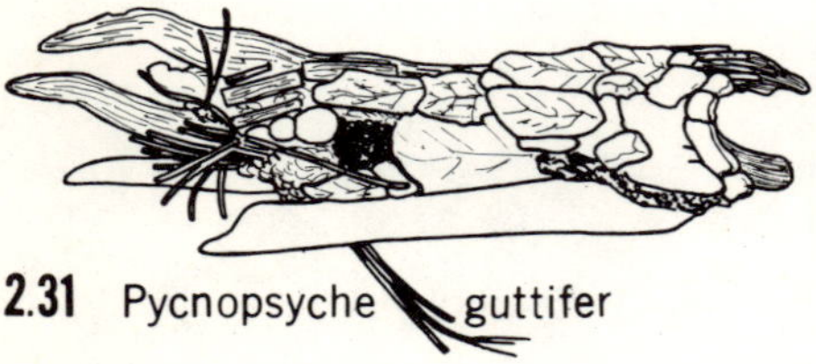

2.31 Pycnopsyche guttifer

2.32 Yphria

30b. Mineral particles concentrated in a V-shaped area on the ventral (concave) side of the case tapering back from anterior margin . . . . (Fig. 2.32, *Yphria*) Kitagamiidae (= Limnocentropidae)

Figures 2.1 through 2.32 are redrawn from the following sources: Betten, 1934, Figs. 2.12, 2.13, 2.14, 2.15; Lloyd, 1921, Figs. 2.7, 2.8, 2.11, 2.12, 2.13, 2.16, 2.17, 2.18, 2.20, 2.21, 2.22, 2.24, 2.25, 2.26, 2.28, 2.30; Noyes, 1914, Figs. 2.2, 2.3, 2.24; Ross, 1938, Fig. 2.5; Ross, 1944, 2.1, 2.5, 2.6, 2.7, 2.10, 2.14, 2.19, 2.27; Wessenberg-Lund, 1943, Figs. 2.2, 2.4, 2.5, 2.8, 2.24; Wiggins, 1960, Figs. 2.11, 2.29, 2.32.

b. Sound Production in Field Crickets

An additional excellent example of the importance of behavior to taxonomy is found in the work of Alexander (1957a, 1957b) on the field crickets (Orthoptera, Gryllidae, *Gryllus*) of the eastern United States. Some genera of Orthoptera contain species groups that are, at present at least, morphologically indistinguishable. In dealing with five species of *Gryllus,* Alexander considered distribution, seasonal life histories and ecology, and song characteristics, and he conducted laboratory experiments on crossing and hybridization. In Figure 2.33, the song patterns of these *Gryllus* species are shown as they appear on audiospectrographs.

The range and usual number of chirps per minute for six species of *Gryllus* are given in Table 2.1 along with some morphological, ecological and distributional features (all data after Alexander, 1957a).

Table 2.1 Characteristics of six species of *Gryllus*

	Chirps/Min.		Morphology				
Species	Range	Usual No.	Total Length, mm.	Ovipositor Length, mm.	Coloring	Habitat	Geographical Distribution
domestica					Yellow to yellowish brown	"House" cricket (European import)	Introduced over most of U.S.
assimilis							Southern Florida
firmus	56–168	100–120	>18	19–23	Brown	Sandy areas	Atlantic and Gulf States
pennsylvanica [a]	120–370	150–240	<23	<18	Brown, pronotum without light spots on sides	Fields	Most abundant and continuously distributed field cricket in northeastern U.S.
vernalis	108–246	180–200	12.8–19	9.5–15.5	Solid black	Deciduous forest, leaf litter	Central eastern U.S.
fultoni	176–360	240–300	12.8–19	9.5–15.5	Black, tegmina brown	Deciduous forest, leaf litter, forest borders, dry hillsides, old fields	Southeastern U.S.
rubens [b]	>*fultoni*		<23	<18	Brown, pronotum with light spots on sides	Fields	Southeastern U.S.

[a] Most frequently encountered species in the North.
[b] Most frequently encountered species in the South.

Materials and Methods. Live field crickets (*Gryllus*) should be collected from a variety of habitats and established in the laboratory (see Appendix for culture methods). Make tape recordings of the male songs; these recordings will be used also in the experiment in Chapter Five, Section C. Satisfactory results for purposes of these experiments can be obtained with an inexpensive machine having two speeds; 7½ and 3¾ inches per second are the preferred speeds but others will serve if necessary. Make the tapes by first recording a metronome sequence to establish accurate time intervals. Recorded at the faster speed, the tape is played back at the slower speed; with a stop watch (corrected by timing metronome intervals at the slower speed and comparing with normal speed intervals) the number of chirps per minute can be determined for the various species at hand.

For detailed information on refined equipment for recording and analyzing insect songs, see Borrer and Reese (1953), Alexander (1957a, 1957b) and Haskell (1961). A long-play record of insect songs is commercially available (Appendix).

Figure 2.33 shows the frequency of cricket chirps. Comparison of taped chirps made in the laboratory with this figure makes it possible to identify the various species shown. Check the identifications against the morphological, ecological and distributional data given in Table 2.1. For a detailed key to the species of *Gryllus* of eastern United States and other pertinent information, see Alexander (1957a), Alexander and Bigelow (1960) and Alexander and Walker (1962).

c. *Fruit Fly Mating*

The mating behavior patterns for a number of species of *Drosophila* are described by Spieth (1952) in sufficient detail that they can be used as criteria for separating adult flies at the species level. Recently Waldron (1964) has shown that audiospectographs of the beating wing of the male show different sound patterns for *D. pseudoobscura* and *D. persimilis*.

Materials and Methods. The materials required, in addition to cultures of various species of fruit flies (see Appendix), for the study of *Drosophila* mating behavior are given in Chapter Five, Section B.

After careful observations on the courtship and mating of as many species of *Drosophila* as possible have been made, it is instructive to attempt the construction of a key to the species at hand, based entirely on observable behavior.

The examples studied in this section—case building in Trichoptera, mating behavior in *Drosophila,* and sound production in the genus *Gryllus* —are but a few of those available in a richly rewarding field of study.

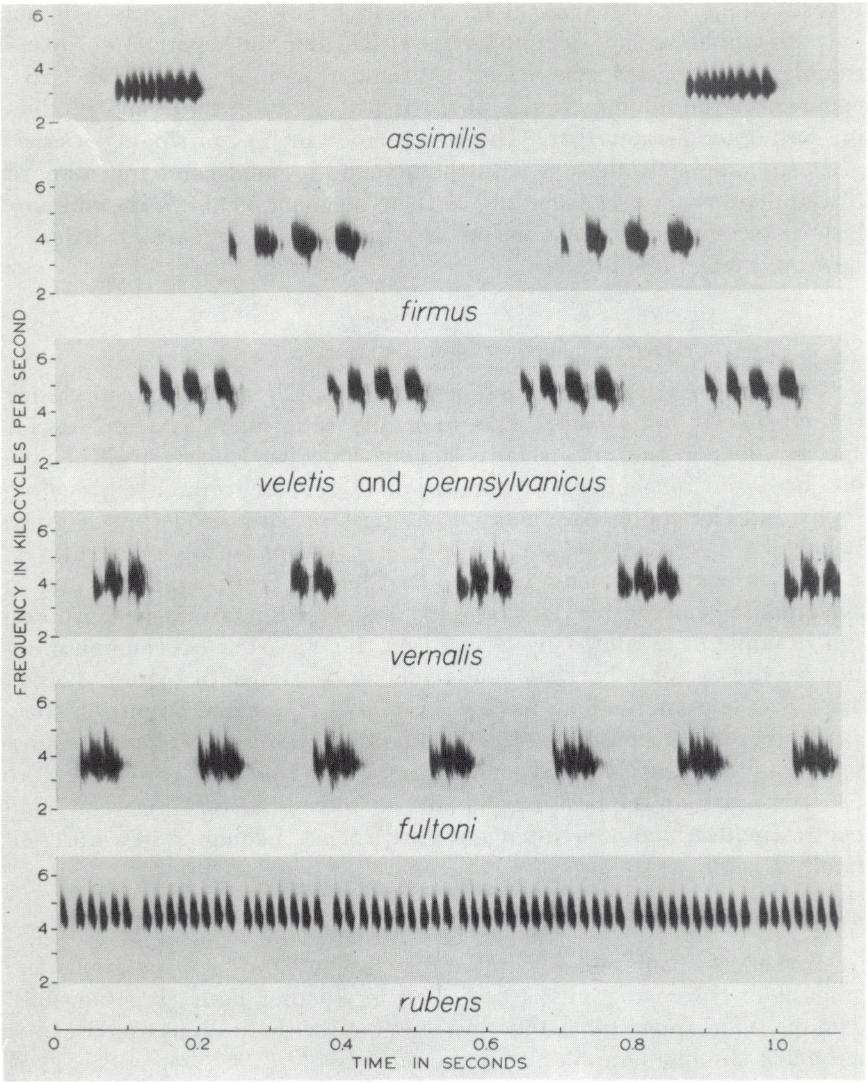

Figure 2.33. Audiospectrographs of the calling songs of the field crickets of eastern North America (*Gryllus* species). *G. veletis* (Alexander and Bigelow) and *G. pennsylvanicus* have identical songs but are seasonally separated. *G. vernalis* (Blatchley) and *G. fultoni* (Alexander) are sibling species but occur in different sorts of woodland. *G. assimilis* (Fabricius) is a neotropical species found in southern Florida. (Previously unpublished plate kindly furnished by Dr. R.D. Alexander)

Investigations of taxonomic differences in behavioral patterns will not only strengthen species definitions but also enable the separation of many troublesome "species complexes." Alexander and Moore (1958: 126), commenting on the importance of singing behavior in the sound producing Orthoptera, stated that, "The whole mode of life [of singing insects] . . . is so completely tied up with the system of sound communication of the adults that an understanding of their behavior, ecology, distribution, history and taxonomic relationships is impossible without a knowledge of how this system operates."

2. PHYSIOLOGICAL IDENTIFICATION

Consistent physiological differences undoubtedly exist between closely related species, but sufficient data to enable using such characteristics for species identification are usually lacking. One particularly fruitful area has been the biochemical identification of species through chromatography and electrophoreses; a simple exercise in chromatography is given below.

The various experiments outlined in Chapter Four could be used to accumulate comparative data on related species for taxonomic purposes. For example, it is interesting to compare the curves of oxygen consumption for the two closely related caddis fly larvae shown in Figures 4.5 and 4.6. On a per gram per hour basis *Pycnopsyche lepida* and *P. guttifer* show strikingly different patterns of oxygen consumption when plotted against temperature or dissolved oxygen concentration. The same two species are also compared on the basis of Q_{10} (see Chapter Four for discussion) and contrast with *Molanna uniophila,* a third species belonging in a different family (Table 4.1).

a. Chromatographic Separation of Drosophila *Species*

MATERIALS. Equipment suitable for chromatographic separations and analysis of chromatograms is described in Chapter Four (see Fig. 4.2). The pie plate apparatus is the most suitable for the present exercise. For details of chromatographic technique see Cassidy (1957) and Block *et al.* (1958); a brief historical treatment of the technique has been presented by Lonert (1964). Procedures especially relevant to the experiment below can be found in Buzzati-Traverso (1960). Two or more species of *Drosophila,* both males and females; newly emerged, unfed adults should be used (see Appendix for sources of *Drosophila* cultures and culture methods). Glass rods and fine forceps are needed. Other insect species may be used as available.

METHODS. If filter paper strips are to be used, make a penciled line across each one about one-half inch above the solvent surface; if the pie plate apparatus is used, draw a small circle at the center of the disc of filter paper. Special care must be taken when handling the filter paper to avoid contamination of its surface with extraneous materials or finger prints. Remove the head, wings and legs from fresh live specimens and squash the body with a glass rod on the penciled line or circle. Develop the strips at 10° C. for four to six hours. When the discs or strips are removed the solvent front is marked with pencil and the papers are dried at room temperature (see Lewallen, 1957). The discs or strips are then checked under an ultraviolet lamp for florescing spots and developed with ninhydrin spray.

Compare the chromatographic patterns of individuals of a given sex, of sexes of a species and of comparable individuals of different species.

This procedure can be expanded so that representatives of different genera, families or orders can be compared. Feeding stages, if used, must be eviscerated before analysis.

CHAPTER THREE

Genetics

The geneticist owes a great deal of his knowledge to insects, for few animals are more extensively used by him than are various insect species, particularly those of the dipteran genus *Drosophila*—so easily cultured in large numbers and having such a short life cycle. The wasp *Habrobracon*, the flour moth *Anagasta*, the grain beetle *Tribolium* and other insects also have helped establish genetic principles.

Many excellent genetics texts are available—Srb and Owen (1952), Sinnott, Dunn and Dobzhansky (1958), and King (1962)—so that detailed discussions of genetic principles are unnecessary here. It is strongly recommended that the student consult any of these references, both for the fundamentals and the details of the subject. The laboratory texts by Strickberger (1962) and Demerec and Kaufmann (1962) are strongly recommended for work with *Drosophila*, and the book by Martin (1947) excellently summarizes knowledge of *Habrobracon* genetics.

In the succeeding exercises only the basic techniques used by geneticists working with insects will be summarized. It must be understood that these exercises barely scratch the surface of available genetic experimentation.

A. GENETIC CROSSES WITH *DROSOPHILA*

It is not the purpose of this book to illustrate all or even the majority of the informative crosses using *Drosophila* that are possible, but no journey through the gamut of experiments using insects would be complete without at least a rudimentary mention of the work that has been done using the little fruit flies. At least half the literature in genetics deals with these insects, and more is known about their genetics than is known of that of any other organism, including man.

Two of the many available crosses have been selected for inclusion here, both dealing with single mutants. Nothing will be done with more than one gene at a time; for such experiments the student is referred to the suggestions put forth by Strickberger (1962). Only autosomal and sex-linked crosses will be considered here.

The time for egg-to-adult development of *Drosophila* is ten to fourteen days, depending primarily on temperature and food conditions; about a month must be allowed for an experiment, since two generations after the original mating are necessary.

One basic statistical test commonly used in genetics is the Chi square test. This is a test of the "goodness of fit" of the observed data to the expected ratios; in other words, it shows what the probability is that the deviation of the observed frequencies from the expected ones is due solely to chance. The technique for the calculation of a Chi square is shown in Table 3.1. The numbers refer respectively to the number of objects in class *A* and the number of objects in class *a*.

Table 3.1 An example of Chi square application

Data		
A sample of 200 individuals falls into 2 classes	*A* and	*a*
The observed frequency (*o*) for each class is	145	55
But the expected frequency (*e*) for each is	150	50
Calculation		
Find the deviation $(d) = (o - e)$ or $(e - o)$	5	5
Find the square of each deviation (d^2)	25	25
Divide each (d^2) by the respective (e), (d^2/e)	0.16	0.5
Chi square is the sum of all (d^2/e) values	0.66	
Chi square table		
Enter with 1 degree of freedom, find probability	between 0.30 and 0.50	
Conclusion		
The odds are between 30 and 50 in 100 that the distribution of the 2 classes in this experiment are due strictly to chance		

Now check the Chi square value in a table of Chi square, such as the one in Sinnott, Dunn and Dobzhansky (1958: 397), with one degree of freedom. Usually the degrees of freedom are one less than the number of classes of observation. The above Chi square value of 0.66 with one degree of freedom carries a probability of between 0.30 and 0.50—that is, the amount of deviation between observation and expectation may be due to chance thirty to fifty times out of a hundred. In biology any probability of greater than 0.05 is considered nonsignificant or arbitrarily due to chance, so it may be said that the above deviation of the observed results from the expected ones was due to chance alone.

The materials for the two crosses are substantially the same, but the methods and the handling of the data are somewhat different.

MATERIALS. Mutant and wild-type *Drosophila* stocks; vials with *Drosophila* food, plugged with cotton; bottles with *Drosophila* food and tight-fitting caps; yeast suspension in water; sterile paper strips; dissecting microscopes; 4 inch by 4 inch pieces of glass plate, metal, or white cardboard for counting flies; fine camel's hair brushes; ether and etherizers.

1. SINGLE GENE AUTOSOMAL CROSS

For this cross a recessive gene will be used for the sake of illustration. In the F_1 generation of such a cross all the flies should be wild-type, but the F_2 progeny will show a three to one ratio of wild-type flies to mutant ones. Such a cross is discussed in Sinnott, Dunn and Dobzhansky (1958: 38–42).

For the purposes of this study the mutant gene *vestigial wing* (vg) of *D. melanogaster* will be considered. Flies homozygous for vestigial (vg/vg) have greatly reduced wings (see Sinnott, Dunn and Dobzhansky, 1958: 40) and are flightless. Flies homozygous for wild-type (vg^+/ vg^+) or heterozygous (vg^+/vg) have fully formed wings and can fly normally, the wild-type allele being dominant over vestigial.

METHOD. Put a folded sterile paper strip into the medium of a vial of *Drosophila* food and add a drop of yeast to the medium. Prepare several such vials.

Lightly etherize both vestigial winged and wild-type flies and look for virgin females of both (Note: To save time these flies may be isolated by the instructor and provided for the class). The ventral surface of the abdomen of a virgin female is not as distended as that of a mated female, and at the proximal end of the ventral surface of the abdomen there is a "dark comma" showing through the integument on the left side. This is a portion of the gut; when a female has mated the ovaries expand and obscure the intestine within a matter of minutes. In addition, epidermal pigment is light in young females. Flies darken as they age during the first twelve hours following emergence from the pupa. Matings ordinarily do not take place in the first twelve hours. Once a female has mated she cannot be used a second time since sperms are stored for several days.

Half the matings should use vestigial winged females with wild-type males; the other half should employ wild-type females and vestigial winged males. These matings should be in separate vials, of course. Be extremely careful not to drop the anesthetized flies into the medium, where they may get stuck and eventually die. Dead flies don't produce progeny! Keep the vials on their sides until the flies have revived, then place the vials in a warm (about 75°–78° F.) place for two weeks. After larvae (F_1) appear, the parents should be removed so they will not be counted as progeny. When the F_1 flies have emerged etherize them lightly and examine them under a dissecting microscope. All the flies should be wild-type; if any vestigial winged flies appear, the parental female was probably not a virgin. Check the F_1 progeny again for virgin females and isolate them. Perform this same procedure for each original mating vial.

Prepare two culture bottles with yeast, paper, and *Drosophila* food as before. Into one of these place the *virgin* females and some vestigial males that the instructor will provide. This is the **test cross** using a mutant male to test the genotype of the F_1 female. We expect that half will be wild-type and half vestigial winged flies in the test cross progeny. Why? Into the other bottle place five or six pairs of F_1 flies (it does not matter in this case whether or not the females are virgin since inbreeding brother with sister is required) to serve as parents for the F_2 generation. Be sure to remove the F_1 adults after one week. We expect three-fourths of the F_2 progeny to be wild-type and one-fourth to be vestigial winged. Why?

In two weeks examine the F_2 and test cross progeny. Test your results with the expected results by a Chi-square test as shown in Table 3.1.

Diagram the cross of vestigial wing by wild-type. Does it matter which original parent was dominant or recessive genotype? Why or why not?

2. SINGLE GENE SEX-LINKED CROSS

Sex-linked characters are those transmitted by the X or Y chromosome, those chromosomes by which the sexes differ. In *Drosophila*, as in man, males have one X and one Y chromosome while females possess two X chromosomes, but this situation is by no means universal in insects (see Fox and Fox, 1964: 215–220). In the scheme exhibited by *Drosophila*, females may be either **homozygous** or **heterozygous** for a trait carried on the X chromosome, whereas males receive only a single "dose" of X chromosome heredity (**hemizygosity**). Since the X chromosome of the male is received initially from his mother, sex-linked traits in *Drosophila* are passed from mother to son.

The X chromosome mutant gene that will be used for this cross is *bar-eye* (B), a semidominant gene which has different expression in the homozygous and heterozygous conditions (see Srb and Owen, 1952: 89).

METHOD. Place yeast and paper in vials as before, isolate virgin females of both bar-eye and wild-type flies and mate virgin females of one to males of the other in separate vials as before. The same precautions about not letting the flies fall into the medium apply.

In two weeks count the F_1 progeny, both as to sex and appearance (**phenotype**). Are the males like their fathers or like their mothers? Why? What characteristics do the females display? Examine the eye shape very carefully. Is the appearance of the progeny the same in a bar-eye female by wild-type male cross as in a wild-type female by bar-eye male cross? Explain.

Prepare yeast and paper in only one culture bottle of *Drosophila* food to provide for an F_2 generation. The F_2 males will automatically test the genotype of the F_1 female. Explain. Place five or six pairs of F_1 flies in the bottle and leave it in a warm place for two weeks. Discard F_1 adults after one week. Count, sex and characterize the F_2 progeny. There should be two classes of males and two classes of females. With this in mind, perform a Chi square analysis of your data on the assumption that all four classes should be equally frequent. There will be *three* degrees of freedom; check the results of your Chi square computation with the tabular results bearing this fact in mind. Do the phenotypes differ in the F_2 generation depending on whether bar-eye or wild-type was used as the original parent female? Why? Diagram a sex-linked cross showing the results of bar-eye female by wild-type male as opposed to wild-type female by bar-eye male in the original cross.

B. DIPTERAN GIANT SALIVARY GLAND CHROMOSOMES

The giant salivary gland chromosomes, figured particularly well by King (1962: facing p. 132), have been useful in the location of mutant sites in *Drosophila*. These huge cells are unusual in that the chromosomes divide many times without concomittant cell division and do not disjoin, resulting in a thick "rope" of chromatids all displaying heavy concentrations of nucleic acid, the bands that define genic loci. The two homologous chromosomes pair as if in meiotic prophase, so that each salivary chromosome arm is actually two chromosomes. Such a chromosome, composed of many chromatin strands, is known as a **polytene** chromosome. In the presence of inversion or deletion differences between pairs of chromosomes, loops are formed which mark the position of such anomalies. Such a situation is illustrated by Sinnott, Dunn and Dobzhansky (1958: 191).

The individual chromosomes are identified by their characteristic bands and puffs at the tips. Once identified the individual chromosome may be traced back to its origin in the chromocenter, the large amorphous center mass, which represents common centromere material of all chromosomes grouped together.

The larger the larva the better the chance of obtaining usable salivary gland preparations; mature larvae should be provided for this experiment. If the larvae of *D. melanogaster* are grown in an abundance of yeast at 15° C., rather than the optimum 25° C., the larval period will be longer, and the resulting larvae will be larger and more suitable for salivary gland preparations.

MATERIALS. Large mature *Drosophila* larvae; dissecting needles, preferably with angled tips; Pringle's solution (see Appendix); slides; no. 2 cover slips; Lacto-acetic-orcein (see Appendix); bibulous paper; dissecting and compound microscopes.

METHOD. Select a large *Drosophila* larva and place it in a drop of Pringle's solution on a slide under the dissecting microscope. Examine the head region of the larva, noting the sclerotized mouthparts.

Place the flat edge of one of the dissecting needles on the middle of the body of the larva and pierce the head, just behind the mouthparts, with the tip of the other needle. Carefully pull the head away from the rest of the body and note the nearly clear salivary glands, outlined by white fat, which follow with the head. Salivary glands are long cylindrical paired structures just posterior to the "brain," and are the sites of the giant chromosomes.

Separate the salivary glands from the head and carefully remove as much extraneous tissue from the salivary glands as possible. Transfer the salivaries to a drop of lacto-acetic-orcein dye on another slide. Let the gland tissue stand for at least five minutes and let it partially dry; then cover with a clean cover slip. Place the slide and cover slip between two pieces of bibulous paper and press down on the cover slip gently but firmly; the paper will remove the excess lacto-acetic-orcein.

Examine the prepared slide under the compound microscope and locate a squash preparation of chromosomes similar to that figured in Sinnott, Dunn and Dobzhansky (1958: 191). Find the free terminal ends of the individual chromosomes and identify them from a figure (provided by instructor) such as that in King (1962: facing p. 132), which is excellent for identifying the chromosomes of *D. melanogaster.* Use the patterns of bands and bulges at the tips of the chromosomes for identification. Draw one or two such chromosomes after you have identified them.

MODIFICATIONS. Polytene chromosomes can also be found in the cells of malpighian tubules, midgut wall and other tissue of *Drosophila.* Many Diptera display giant polytene chromosomes, not all located in the salivary glands of the larvae. Whitten (1964) records giant chromosomes in the pupal post-tarsal pulvilli of cyclorrhaphan flies, particularly the flesh fly, *Sarcophaga*. The pupae are dissected early in their development, the pulvilli removed, squashed and stained with lacto-acetic-orcein, as was done above with *Drosophila*. A representative photograph of the giant chromosomes from the hypodermis of the sarcophagan pupal pulvillus is given by Whitten (1964: 1438).

The use of larval giant salivary gland chromosomes to separate sibling species of black flies (Diptera: Simuliidae) is summarized by Rothfels (1958). Particularly informative papers on the use of this method of unraveling sibling species groups are those of Dunbar (1958), dealing with the insects previously recorded as *Eusimulium aureum* (Fries), and of Barsur and Rothfels (1959) on the *Cnephia mutata* (Malloch) complex, both of which suggest that several species may be lumped as a single taxonomic entity as now defined. The use of chromosomal analyses in insect systematics is still in its infancy, but it already holds the promise of solving many previously baffling problems.

Rather than lacto-acetic-orcein, Rothfels and Dunbar (1953) discuss a technique using Feulgen stain in the preparation of these chromosomes. Fast green may also be employed, but it is of more limited value than is either the Feulgen or lacto-acetic-orcein stains.

If simuliid larvae are available it will be informative to repeat with them the experiment outlined above for *Drosophila*. Although the full distribution throughout the insects of these chromosomes is not known, it appears that the majority of the nematoceran Diptera may possess them in their larval stages.

C. SEX DETERMINATION IN *HABROBRACON*

Habrobracon juglandis (Ashmead) has been a standard laboratory animal for genetics since Whiting (1918) first recorded its peculiar mode of sex determination—haploid males and diploid females. Martin (1947) summarizes most of what is known about the biology, culture, sex determination and genetics of this interesting little insect.

The most reliable method of separating male *Habrobracon* from females is to count the antennal segments: males have twenty to twenty-two segments and females thirteen to fifteen (Fig. 3.1). Other differences between the sexes, less evident and not considered here, are detailed by Martin (1947: 6–11).

Whereas the sex determinants in *Drosophila*, man and most other organisms are directly related to the presence of sex chromosomes, sex determination in *Habrobracon* is dependent upon whether the organism is hapoid or diploid. All the chromosomes of *Habrobracon* are considered autosomes. An unfertilized female *Habrobracon* will still produce progeny, all haploid males. Female offspring are produced only by the union of male and female gametes, but not *all* the progeny of a mated female will be females: at the time of fertilization about half the eggs will be held back and not fertilized and the progeny of a mated female, therefore, will be about half males and half females. This type of sex determination with the parthenogenetic production of males (**hemizygoid parthenogenesis**) is discussed by Fox and Fox (1964: 219).

Habrobracon is a parasitic braconid wasp which feeds in the larval stage on certain caterpillars and it is necessary, therefore, to provide caterpillars to keep a *Habrobracon* colony. The host moth of choice is *Anagasta kuhniella* (Zeller),[1] a pyralid moth, itself the subject of many genetics papers. Directions for keeping an *Anagasta* culture are given in the Appendix.

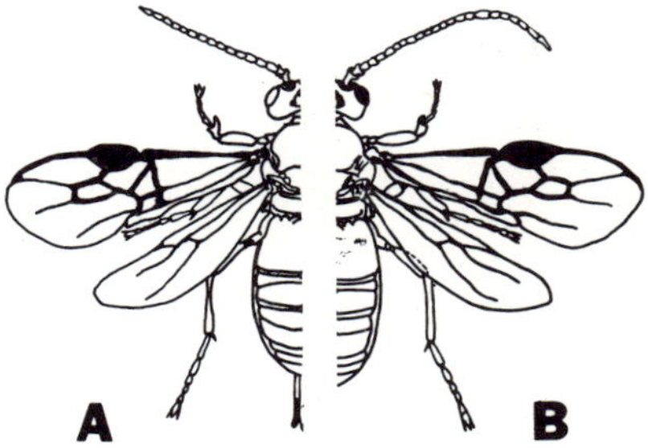

Figure 3.1. Dorsal view of *Habrobracon juglandis* Ashmead. (A) female, (B) male. (Martin, 1947)

The female *Habrobracon* will sting and paralyze an *Anagasta* caterpillar; the eggs are then laid on the caterpillar and hatch larvae in about two days. The larvae affix themselves to the integument of the host, feed on the body juices and pupate after about two days. The entire egg to adult cycle takes about ten days (Martin, 1947: 12–17).

MATERIALS. Males and virgin females of *Habrobracon* (it may be informative to have mutant females in order to show the source of the chromosomes in the males); *Anagasta* caterpillars; shell vials with cotton plugs; dissecting microscopes; ether and etherizers; constant temperature cabinet.

METHOD. Place an *Anagasta* caterpillar in each of several shell vials. Etherize *Habrobracon* virgin females (Note: these wasps are far less susceptible to ether than are *Drosophila* and a longer exposure to ether is necessary) and place one in each shell vial. Etherize *Habrobracon* males and place one in each of half the shell vials.

Observe the mating of the wasps and, if time permits, the stinging of the caterpillars and oviposition of the eggs.

[1] Heinrich (1956: 186, 299–300) erected *Anagasta* for this species, removing it from its familiar placement in *Ephestia* for biological reasons.

Place the shell vials in the constant temperature cabinet preset to 30° C.; ideally the humidity should be about 80%, but is less critical than temperature.

After three days remove the adult wasps, which may be kept alive for several weeks in shell vials if fed a honey and water mixture and thus kept for later experiments. Males will feed only on plant juices, but the females supplement their diets with the body fluids of caterpillars when available.

In ten days to two weeks examine the progeny from each of the vials. Are all the progeny of virgin females male wasps? What is the sex ratio of the progeny of the mated females?

If desired, a mutant character may be used in the female to show the source of the F_1 male chromosomes. In such an experiment an F_2 generation, using virgin F_1 females, is desirable to test the genotypes of the F_1 females. Various mutants that may be used for this type experiment are suggested by Martin (1947).

C H A P T E R   F O U R

Physiology

Since a number of excellent texts dealing with basic insect physiology are available (Wigglesworth, 1956; Roeder, 1953; Campbell, 1959; and Patton, 1963), there would be little point in attempting here a review of this extremely broad and exciting field. Rather, the exercises in the present chapter sample some of the currently active research areas.

Because apparatus for the measurement of physiological phenomena is often highly specialized and expensive, most sections in this chapter include both introductory and advanced exercises. It must be emphasized that structure and function cannot be divorced; therefore the worker must be familiar with the morphology of the insects studied, especially with the neural anatomy. For this reason the student is directed toward a review of the first chapter of this book and Chapters Two through Six in Fox and Fox (1964).

For additional experiments concerned with insect physiology the reader can consult Kalmus (1960) and Welsh and Smith (1960).

A. SENSORY RECEPTION

The experiments outlined below are really in the realm of behavioral physiology (or physiological behavior, depending upon one's point of view). A general treatment of the structure and function of sense receptors of insects and arachnids is given in Fox and Fox (1964: Chapter Six).

As Dethier and Chadwick (1948) and Dethier (1953, 1956) have indicated, there is reasonably good evidence, at least in terrestrial species, for three separate chemical senses: a common chemical sense, an olfactory sense and a gustatory sense. Of course, functional interrelationships between the three sensory systems in live, intact animals often makes it difficult to distinguish the three types.

The common chemical sense includes responses to relatively high concentrations of strong irritants such as ammonia. Responses to irritants are covered in part in the section below concerning metameric independence.

The olfactory receptors are extremely sensitive, responding to as little as several molecules of a volatile compound at normal temperatures. An outstanding example, sexual attraction in saturniid moths, has been utilized as the first exercise.

Gustatory receptors typically respond to higher concentrations of materials in liquid phase. The degree of sensitivity is exemplified by the work with taste responses in adult dipterans, while selectivity for food plants, mediated through both smell and taste receptors, can be demonstrated with lepidopteran larvae.

For interesting reviews of chemical insect attractants and chemoreception the reader should refer to Jacobson and Beroza (1963, 1964) and Hodgson (1958).

1. SEXUAL ATTRACTION IN SATURNIID MOTHS

The attraction to males of some insects exerted by virgin females has long been familiar to naturalists; that the attraction is primarily chemical has been known for over fifty years. Scent organs with odorous sexual attractants are perhaps best developed in lepidopterous families Bombycidae, Lasiocampidae and Saturniidae—moths with short adult life expectancies and ones in which fully developed eggs are available for fertilization immediately after adult eclosion (Dethier, 1947: 21–22).

The female moth assumes a "calling pose" with the attractant gland, located near the female genitalia, everted. This gland secretes the attractant, which is then dispersed through the air. That this secretion is the attractant is attested to by males attempting to copulate with pieces of blotting paper wetted with the secretion (Dethier, 1947: 22). In Saturniidae, copulation will also occur with abdomens clipped from females during the calling period (unpublished observation, LDM).

Surprisingly low concentrations of the attractant are effective: marked males of the Indo-Australian saturniid *Actias selene* (a relative of our luna moth) were recovered from a distance of 11 km., and marked males of the gypsy moth *Lymantria dispar* have been recovered 3.8 km. from the point of release (Dethier, 1947: 23; Carthy, 1958: 276). One of the authors (LDM) has recovered males of the promethea moth, *Hyalophora (Callosamia) promethea*, from a distance of 3 km.

The chemical nature of the female sexual attractant of saturniid moths is still unknown, but that of the gypsy moth has been isolated (Haller *et al.*, 1944) and is presently being used in control programs by the U.S. Department of Agriculture. In addition, the sex attractant of the cockroach has been isolated by Jacobson *et al.* (1963).

Theoretically these attractants should be species-specific, but this is only true to a limited extent (Carthy, 1958: 276–277). Members of the genus *Hyalophora* will readily hybridize, but since the F_1 females are sterile, sympatric species remain distinct (Sweadner, 1937). Intergeneric copulations have also been reported, but are almost always complete failures (Collins and Weast, 1961).

For the present exercises saturniid moths will be used, not only for their large size and easy manipulation, but also as a source of material for juvenile hormone extraction in a later exercise. For convenience, the promethea moth, which mates in the late afternoon, is the moth of choice. The cecropia moth *Hyalophora* (*Hyalophora*) *cecropia* would be preferable, but its mating activity is highest only near dawn (see Table 6.1).

MATERIALS. Virgin female saturniid moths; wire cages; hand lens; blotting paper or filter paper; scissors.

METHODS. Examine the ventral portion of the abdomen of a quiescent female and observe especially the structures near the tip. A drawing made now will be helpful for later comparison.

Place the female in a wire cage about the time of the onset of mating activity; the attraction times for some common species are shown in Table 6.1. Place the cage in the open near a wooded area where the species should occur. Watch the female periodically, particularly the tip of the abdomen, for mating activity. When the female assumes her "calling pose" she extends her abdomen and protrudes the area around the female genitalia. Examine the tip of the abdomen of a "calling" female with the hand lens and locate the scent gland.

The caged female "calling" in this manner serves as a "trap" for males in the vicinity. As males are attracted, capture them and place them in a cage *separated* from the female. These males will be used in later experiments of this exercise and may also be used for the extraction of juvenile hormone (Section F 1 in this chapter).

Now touch a small piece of blotting or filter paper to the scent gland at the tip of the female abdomen. The moisture on the paper is the attractant substance. Place the paper in the cage with the males and observe their actions. Do they become agitated? Are they attracted to the blotter? Do they attempt copulation with it?

Cut the antennae from a male and place him near a "calling" female. Note his reactions. Is he now attracted to her? This demonstrates that olfaction is primarily centered in antennal receptors.

When a female has assumed her "calling pose," clip off the tip of her abdomen. Carefully, so as not to spread the attractant around too much, place this abdomen in a cage with some males and observe the males' responses. Are they attracted? Will they copulate with this free abdomen? Repeat this test with the head, thorax, and wings of the female. This experiment localizes *all* the chemoattraction to the abdominal region of the female.

A population capture and recapture technique utilizing the chemoattraction of male saturniids to virgin females is given in Chapter Six.

2. LOCALIZATION AND NATURE OF TASTE RESPONSES IN ADULT DIPTERA

The taste receptors of most insects, including Diptera, are located in the pulvilli of the post-tarsal segment (Minnich, 1926). The following experiment demonstrating taste responses in Diptera is essentially taken from the work of Dethier (1953c), who not only considered the physiology of taste responses in the blowfly, *Phormia,* but also showed the histology of the chemoreceptors.

The method of mounting the living flies for study is substantially that of Frings (1946).

MATERIALS. Any of the following adult flies are suitable (in descending order of size): *Sarcophaga* (flesh fly), *Phormia* (blowfly), *Musca* (housefly) or *Drosophila* (fruit fly); glass rods or wooden applicator sticks; block with holes drilled in it to accept the above rods; paraffin or wax; glucose, sucrose and/or fructose in solutions of 0.005, 0.01, 0.05, 0.1, 0.5 and 1.0*M;* other chemicals, such as solutions of salts, acids, bases and alcohols; distilled water; watch glasses to hold test solutions, or applicator sticks with cotton swabs; anesthetic, either CO_2 or ice.

METHODS. Anesthetize some of the flies, either by exposure to CO_2 gas or by hypothermia (treatment with ice) as suggested by Gross and Halpern (1961), a day or so before the actual experimenting is to take place.

Prepare the wooden applicator sticks or glass rods by placing a small amount of wax on the tip of each. Heat the wax to tackiness and touch it to the mesonotum of one of the flies; when the wax cools, the fly will be affixed to the rod. Discard all injured flies and use only those that are healthy. Place the rods in the holes in the block for safekeeping until the time of the experiment.

Starve the flies for twenty-four hours before the experiment, but provide them with enough distilled water to compensate for natural water loss. Starvation, of course, makes the flies more willing subjects for the experimentation that follows. Flies should be tested with distilled water before proceeding with an experiment; those showing a positive response should be discarded.

A **positive response** to stimuli applied to the tarsi is defined as a lowering of the proboscis toward the source of the stimulus, here the watch glass or cotton swab; a **negative response** is manifested by the lifting of the tarsi away from the stimulus (avoidance).

Holding a fly by its attached rod, dip the tarsi into a 0.005*M* sugar solution in a watch glass. Is there a positive response, or not? Repeat the process successively with 0.01, 0.05, 0.1, 0.5 and 0.1*M* sugar solutions and record the first positive response, washing the tarsi with distilled water between trials. This experiment should be repeated with several flies and data compared to determine the lowest concentration of sugar that will elicit a positive response—that is, the **threshold concentration** for this sugar. Alternatively a cotton swab, dipped in a test solution, can be rubbed on the tarsi. Are the threshold concentrations the same for all sugars tested? If time permits a more accurate estimate of the threshold concentration may be found by testing various concentrations of sugar between the apparent threshold and the highest concentration of sugar at which no response was obtained.

Repeat this experiment using solutions of salts, acids, bases and alcohols. What responses are noted to these materials? Find the threshold values, if any, of some of these chemicals.

Mix subthreshold concentrations of two or more of the sugars and introduce this mixture to the tarsi of a fly. Is there a response? Apply the same solution to the tarsi several times in rapid succession. Is there now a response? This is called a **subliminal response.**

Now, mix an above-threshold concentration of a sugar and add to a solution of another chemical that previously failed to elicit a positive response. Introduce this mixture to the tarsi of a fly. Is the response positive (controlled by the response to sugar) or negative (avoidance response to the chemical)? Repeat with various concentrations of the other chemical in the above-threshold sugar solution.

MODIFICATIONS. Electrophysiological studies have been performed on taste reception and reaction of Diptera (Hodgson, 1956, 1958) using various apparatus that may or may not be available in the laboratory. These papers should be consulted in any event; if some of the apparatus mentioned is available, it may be used to provide more refined data on the nature of chemoreception than is derived by the preceding experiment.

3. CHEMICAL FACTORS DETERMINING FOOD PLANT SELECTION BY LEPIDOPTEROUS LARVAE

Relatively few studies have been made on the factors determining food plant selection by caterpillars; the most notable papers are those of Dethier (1937, 1941) and form the basis for the experiments following.

Three separate but interrelated experiments will be performed as parts of this exercise. The first is partially behavioral, the second purely physiological and the last is in part anatomical.

a. Detection of Odor of Food Plant

The following is substantially the experiment designed and reported by Dethier (1937: 18–20) utilizing caterpillars of the familiar monarch butterfly, *Danaus plexippus* (Linné).

MATERIALS. Larvae of the monarch butterfly, *Danaus plexippus* (Linné); fresh leaves of milkweed plants (*Asclepias*); fresh leaves of other plants, such as oak, cherry, grass; cheap perfume; turpentine; large plastic boxes; rectangles of screen wire cut to the size of the bottom of the plastic boxes; brushes.

METHODS. As a control, note first that monarch caterpillars will eat leaves of milkweed plants but will ignore those of other plants.

Cut squares of milkweed leaves and other leaves and place them in the bottom of one of the plastic pans. Introduce caterpillars into the pans and check one-half hour later. Have the caterpillars come to rest on the squares of the milkweed leaves, or have they come to rest at random? Since the leaves are no longer distinguishable by shape, what is suggested as to the nature of food plant selection?

Repeat the experiment outlined above, using leaves coated with turpentine or cheap perfume. The lack of choice apparent in this section of the experiment suggests that food plant recognition is chemical in nature.

Place milkweed leaves and "foreign" leaves in replicate boxes. Place the screen wire over the leaves, making sure no leaves protrude above the level of the screening. Sketch the position of the leaves. Introduce a single caterpillar onto the screen and watch the path of its movement. When it comes to the part of the screen over the leaves, does it move along in a straight line, or does it cast about, trying to orient itself to the leaf? Trace the path of the caterpillar on your sketch. Is there a difference between the actions of caterpillars in cages containing milkweed leaves and those with the foreign leaves? Why?

A white milky latex oozes from the milkweed plants' broken surfaces. Coat this latex on foreign leaves with a fine brush. Place the leaves on the bottom of a pan, cover with screening and introduce a caterpillar into the pan. Do the actions of the caterpillar more closely resemble those of the caterpillars that were caged with milkweed leaves, or those with the foreign leaves? What does this reveal about the nature of the substance guiding food plant selection by monarch caterpillars?

Dip a milkweed leaf into methyl alcohol and present it wet to the caterpillar. Will it eat the leaf immediately? Will the caterpillar eat the leaf after the methyl alcohol has evaporated? Why?

Thus far the effect of olfaction on food plant selection has been shown. To demonstrate the interrelationship between olfaction and taste, coat a milkweed leaf with sodium chloride solution and allow it to dry. Will the caterpillar begin to eat the leaf? Will he continue to eat it? Does this indicate that the centers of olfaction and taste are the same? The experiment should show that the "right" odor can produce a stimulus which may be counteracted by the "wrong" taste.

b. Attraction to Volatile Oils

This experiment is based on work done by Dethier (1941).

MATERIALS. Larvae of the black swallowtail, *Papilio asterius* (Cramer); filter paper; carrot oil; oil of caraway (carvone); oil of anise (methyl chavicol); oil of coriander (coriandrol); thymol; benzaldehyde.

The chemical formulae of most of the above chemicals are given by Dethier (1941: 68).

METHODS. The black swallowtail larva feeds on a number of umbelliferous plants, both cultivated and wild, including carrot, caraway, anise, coriander and dill. The characteristic odors of the first four plants are caused by the respective presence of the first four oils listed in the materials section above.

Dip pieces of filter paper into the six listed oils and allow them to dry *several days* before using. The response to the oils is facilitated by slightly wetting the filter paper with distilled water just before using.

Present pieces of the filter paper treated with each of the oils to larvae and note whether or not they are attracted to them. Dethier (1941: 64) gives the following order of preference: methyl chavicol, carrot, carvone and coriandrol; the other two compounds are rejected.

From the data presented it should be obvious that the larvae are not attracted to particular chemical families of compounds, but rather to *specific* compounds. It is interesting to note that some of the oils, such as methyl chavicol and carvone, are ring compounds, and others, such as coriandrol, are straight-chain compounds.

c. Localization of Olfactory and Taste Centers

This experiment is modified from the study by Dethier (1937: 9–18). He found that larvae responded to taste and to odor only when the stimuli were applied to the head region.

The response of a caterpillar to a disagreeable odor is characteristic: the labrum is retracted, the maxillae and labium are extended downward and forward, and the mandibles are opened; the process is then reversed, beginning with the closing of the mandibles, and a rather violent "spitting" action results.

The taste response evoked by the introduction of sucrose to the mouthparts is a greedy drinking, whereas contact with dilute HCl causes a violent spitting action by the mouthparts.

Materials. Larvae (those used in the preceding two exercises will serve); "caterpillar boards"—pieces of cardboard with bits of thread for tying the larvae down during the experiment; electric cauterizing needle; glass jars of known volume; small pipets; 1*M* sucrous solution; 0.05*M* HCl; ethyl alcohol.

Testing for Olfactory Sensitivity. Place enough ethyl alcohol in the jars to make the following concentrations of alcohol per liter of air: 10, 50, 100 and 200 mg. Seal the bottles and shake well to distribute the alcohol to all surfaces; allow the bottles to stand for a day. The alcohol should now be evaporated and spread fairly evenly throughout the atmospheres within the bottles.

Tie the caterpillars to the caterpillar boards by a loop of thread around the neck and another around the sixth or seventh abdominal segment.

Unseal the jars and quickly lower the caterpillar on the cardboard into the jar, being careful not to disturb the air within the jar. Reseal the jar. After five minutes observe the results. If sensitive to the alcohol the caterpillars will react by spitting. The sensitivity of man to ethyl alcohol is at about 5 to 6 mg. per liter; how does the sensitivity of the caterpillar compare with this? A more accurate estimate of the threshold concentration of ethyl alcohol may be made, if time permits, by making successive tests from the lowest successful concentration, dropping the concentration 5 to 10 mg. per liter for each test.

Perform the following operations with the cauterizing needle under the dissecting microscope on separate caterpillars: (1) remove the third segments of the antennae, (2) remove the third segments of the maxillary palpi, and (3) remove both the third antennal segments *and* the third segments of the maxillary palpi. Before using these animals for experimental purposes, allow twelve to twenty-four hours for the wounds to heal.

Test all the caterpillars in an atmosphere with more than the threshold concentration of ethyl alcohol and observe their responses, or lack thereof, to the alcohol. What does this reveal about the seat of olfaction? Is it dependent upon a single set of sensory receptors, or are the receptors on either the palpi or the antennae enough to evoke the response?

Testing for Centers of Taste Reception. With the caterpillars tied on the caterpillar boards, place a small drop of sucrose solution on the mouthparts with a pipet and observe the response. Repeat the process later with the dilute HCl. Test the caterpillar that had both the palpi and maxillae removed. Is the site of taste the same as that of olfaction?

Perform the following cauterizations with the needle under the dissecting microscope on different caterpillars: (1) remove the epipharynx (the inner part of the labrum), (2) remove the hypopharynx, and (3) remove both the epipharynx *and* the hypopharynx. Again, allow a day or so for the wounds to heal. Since these dissections are difficult, beginning students may find it more satisfactory to coat the structures with wax instead of removing them.

Test each caterpillar again with the sucrose solution and the dilute HCl as in the first part of this exercise. Note the responses, if any, and their intensities, using the control results for comparison. Is taste an all-or-none response (one which is either stimulating to its maximum or not at all), or is there a graded response, based on how much sensory tissue is present?

4. TEMPERATURE AND HUMIDITY RESPONSES

For terrestrial arthropods in particular, maintenance of acceptable levels of temperature and humidity is of prime importance for survival. As would be expected, highly specialized receptors, as well as more generalized sensitivity, coupled with specific behavior patterns are encountered in insects and arachnids.

As in the previous and following sections dealing with sense reception, the exercises below are essentially studies in physiological behavior. Responses of the tested insects to temperature and humidity gradients will demonstrate the presence and location of the suitable receptors and serve to illustrate that insects differ markedly in type and degree of response elicited. Receptors sensitive to temperature and/or humidity are described in Snodgrass (1935), Comstock (1948) and Carthy (1958); Chapter Six of Fox and Fox (1964) should be consulted.

The identification of specific receptors concerned with temperature perception has proved to be most difficult. Many receptors are normally responsive to more than one class of stimulus and receptors sensitive to *both* humidity and temperature are probably the rule. Furthermore, the direct physical relationship between temperature and humidity renders separate laboratory tests difficult and suggests that simultaneous perception in insects would logically be expected in nature.

MATERIALS. The temperature and humidity gradient chamber shown in Figure 4.1 represents only one of many possible designs (see Gunn and Kennedy, 1936). The system shown will permit the evaluation of relative temperature and humidity responses in sizable groups composed of individuals of the same insect or arachnid species. For more detailed studies, additional temperature and humidity measuring devices would be required (see Appendix).

The apparatus consists of a set of six plexiglass tubes each 4 to 6 inches long with an inside diameter of about 1 inch. As shown in Figure 4.1, a delivery tube (1) of smaller diameter than the others fits snugly into a hole in the side of a center tube (2). Lateral (3) and terminal (4) tubes are affixed to the center tube with masking tape, which can be applied and removed easily. The system is sealed with three stoppers,

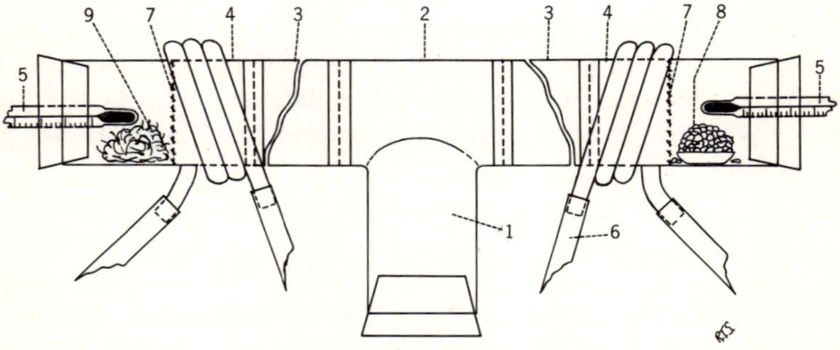

Figure 4.1. Temperature and humidity gradient chamber. 1, delivery tube; 2, center tube; 3, lateral tube; 4, terminal tube; 5, centigrade thermometer; 6, copper or polyethylene tubing; 7, screen; 8, desiccant; 9, wet cotton.

the ones in the terminal tubes being fitted with centigrade thermometers (5). Temperature extremes are established by running hot or cold tap water through copper tubing (6) wound around each terminal tube. In experiments involving light orientation (see exercise below), plastic tubing should replace the copper and appropriate longer periods allowed for the establishment of a temperature gradient. Each terminal tube is fitted with a piece of fine aluminum or plastic screen (7) behind which is placed either a coarse granular desiccant (8) or water-saturated cotton (9) in order to establish a humidity gradient. The entire apparatus can be set on a laboratory table, if the sections are leveled with small supports (additional supports are required under the thermometers). The tubes should be marked with a glass-marking pencil in centimeters along the length of tubes between the retaining screens to provide convenient recording of insect positions.

Insects: *Tribolium confusum* and/or *T. castaneum,* adult mosquitoes (*e.g.*, *Culex pipiens*) and midges (*e.g.*, *Tendipes tentans*) are desirable subjects, along with other species as available.

METHODS. A series of initial experiments should be performed to establish the temperature preferences of each species, keeping the humidity constant, followed by tests with the temperature constant but moisture conditions differing at each end of the apparatus. Finally, various combinations of the two parameters can be tested.

All experiments should be conducted under uniform light conditions; total darkness for periods of twelve to twenty-four hours is best. Periodic temperature readings and animal counts during the experiment should be made.

Each experiment is initiated by first introducing ten to thirty individuals of a given species into the delivery tube; both ends can be stoppered temporarily. Open one end of the delivery tube and insert it into the center section. (Tube 1 should be tilted slightly so that it slopes down to tube 2.) This establishes an initial "T" choice for the insects.

At the end of the experimental period record the positions of the insects; each tube section can be removed and stoppered prior to or following counting as necessary. A hand-tally is a convenient accessory for enumeration. Plot the results on graph paper with number of individuals as the ordinate and a scale in centimeters running from the conditions (*i.e.*, temperature and/or humidity) at the left to those on the right as the abscissa. Assuming a fairly uniform gradient of conditions dry to moist and/or cold to hot, frequency distributions can be evaluated to establish the preferences of the species tested.

As Roth and Willis (1951a, 1951b) demonstrated, various species of beetles, including *Tribolium confusum* and *T. castaneum*, possess humidity receptors (**hygroreceptors**) on the terminal antennal segments and on the maxillary palpi (see also, Pielou and Gunn, 1940; Pielou, 1940; Gunn and Pielou, 1940). Remove the antennae from individuals of the same *Tribolium* species, repeat the experiments and compare the results. Similar series should be conducted using beetles with only the maxillary palpi removed and with antennae and palpi *both* excised.

Temperature perception has been shown to be of primary importance to females of many species of mosquitoes (Culicidae) in locating sources of blood meals (Bates, 1949). The midges (Chironomidae = Tendipedidae), which have significantly different feeding habits, are convenient for comparison. After the responses of representative species of each family have been determined, run several series with individuals without antennae and/or mouthparts. Evaluate the data for possible localization of the receptors involved.

5. PHOTORECEPTION

The physiology of photoreception and behavior associated with it in insects has received an immense amount of research attention (see reviews by Dethier, 1953; Carthy, 1958, Chapters 3–5; Harker, 1961; deWilde, 1962; Jander, 1963). Only experiments at an introductory level are included in this section and in Chapter Five. Several intermediate level experiments are to be found in Welsh and Smith (1960). The advanced researcher can direct himself to such tasks as measuring spectral absorption of extracted insect visual pigment and comparing the derived curves with action spectra as measured behaviorally or electrophysiologically (see Dethier, 1953).

MATERIALS AND METHODS. The materials and methods are those required for the previous exercise on temperature and humidity response, the only addition being a small fluorescent lamp of the type used with dissecting microscopes. This lamp is placed at one end of the previously described gradient apparatus shown in Figure 4.1, and the entire system must be used in an otherwise darkened room. Position the lamp so that reflection effects are minimized.

The experimenter should wait until all animals have moved from the delivery tube (1) into the center section (2); tapping may be necessary to hurry the animals along.

The responses to light, either photonegative or photopositive, are graphed as previously described and should be tested initially with both temperature and humidity kept constant. Following the establishment of responses for several species, the results of experiments with various combinations of the three variables may be evaluated in an attempt to determine the relative importance of each in the over-all orientation of the species under study.

Additional experiments employing species with both compound eyes and ocelli are of interest. By alternately covering the eyes and then the ocelli (see Chapter Five), the importance of these two light receptor systems to orientation can be evaluated. It is interesting to note that dermal receptors are suspected in the apparently eyeless larvae of some insects, although supposed receptor organs composed of retinula-like cells have been found only in the head hypodermi of blowflies (Patton, 1963). Other fairly easy experiments are possible, relating photic response to habitat preference (Lyman, 1945) and orientation (Fraenkel and Gunn, 1940: Chapter 6).

B. BLOOD

Two exercises are included in this section. The first serves to introduce the student to the histology of insect blood cells, the second provides an elementary analysis of the hemolymph. Both exercises may be performed with any insect, arachnid or myriapod, either singly or on a comparative basis.

1. HEMOCYTES

Hemocytes vary histologically among various groups of terrestrial arthropods, as does the relative frequency of each type. Furthermore, cell counts vary with the physiological condition of the individual, apparently correlated to some degree with the point reached in the moulting cycle and with the condition of food reserves (see Fox and Fox, 1964: 157–161).

Many techniques have been used in studies of blood cells, but the one presented below (Sarkarig *et al.*, 1951) has the advantage of being simple and rapid. Alternatively, preparations of blood may be air-dried and heat-fixed and stained with Wright's blood stain. Detailed reviews of histological techniques, as well as aspects of embryology, morphology, physiology and pathology of insect hemolymph and hemocytes are given by Buck (1953), Munson (1953), Jones (1962), Rizki (1962) and Wittig (1962).

MATERIALS. Cockroaches, spiders, millipeds, beetles or other species as available or desired; slides and cover slips; compound microscopes; Sarkarig's stain (see Appendix for formulation).

METHOD. Place a drop of Sarkarig's stain on a cover slip. Obtain a drop of hemolymph by clipping an antenna or the femur of a leg and introduce the hemolymph directly into the medium. Stir with a very thin glass rod and mount by inverting the cover slip on a microscope slide. Petroleum jelly may be used to seal the edges and will preserve the preparation for a few days. Unstained wet mounts may also be used for demonstration, especially if a phase microscope is available.

Note that the above procedure does not require that the animal be sacrificed, so that successive analyses of kinds and numbers of hemocytes may be made. This makes possible many experiments which the student can design to demonstrate hemocyte changes in response to various extrinsic factors. A figure of some of the various cell types is given in Fox and Fox (1964: 158, Fig. 5.7).

2. QUALITATIVE ANALYSIS OF FREE AMINO ACIDS IN HEMOLYMPH BY PAPER CHROMATOGRAPHY

The technique of paper chromatography, which has become such a useful tool for the biologist, may be easily applied to studies of insect hemolymph. This process makes possible the separation and identification of various components of insect blood. Perhaps the simplest analysis is a test for free amino acids; however, it must be remembered that other constituents of hemolymph, such as sugars, also may be separated and identified with suitable detection methods. For techniques and methods see Block *et al.* (1958), Dawson *et al.* (1962), Pant and Agrawol (1964).

MATERIALS. Hemolymph (see preceding exercise for collection methods); solvent—butanol, glacial acetic acid, and distilled water in a 25:6:25 ratio; solutions of known amino acids, 0.5 mg. per ml. each; ninhydrin spray—0.2% in 95% acetone or commercial spray (see Appendix); chromatography chamber; Watman no. 1 filter paper 24 cm. circular and 1½ inch strips; fine drawn glass pipets for spotting.

Various pieces of suitable apparatus for chromatography are shown in Figure 4.2. Large test tubes or graduated cylinders (1 or 2 liter soil testing type without lip) make adequate chambers for ascending chromatography if fitted with appropriately prepared stoppers. Stoppers are to be split in half (Fig. 4.2A) so that the ends of filter paper strips can be held between the halves when the reassembled stopper is inserted into the tube or cylinder (Fig. 4.2B). Alternately, a bent pin can be utilized to hang the filter paper(s) from the lower surface of the stopper (Fig. 4.2C). The filter paper should be cut into strips of suitable length so that they extend into the solvent in the vessel. This paper should not touch the chamber sides (or other papers, if more than one is used in the same chamber). A *pencil* line is used to mark the point, which should be about ½ inch above the solvent level, where the substance is spotted.

Perhaps the most convenient and accurate technique involves the use of two identical pyrex pie plates to form a circular chamber (Figure 4.2D, E). The lips of the plates may be ground to insure airtightness. The following procedure, pertaining to this circular equipment may be easily modified for the ascending type.

METHOD. Find the exact center of a circular piece of filter paper (Watman no. 1, 24 cm.) and in pencil draw a concentric circle about 1½ inch in diameter. Make a hole through the center with a sharp instrument about ¼ inch in diameter to house the wick, which is simply a coiled strip of filter paper. Do not insert the wick until the paper is ready to be placed in the chamber.

Spot minute quantities of hemolymph and of the known amino acid solutions on the circle drawn about the center. Try to keep the spots as small as possible by applying a very small quantity through a very thin pipet or capillary tube, drying, and then adding more until 10 to 50 lambda ($1\ \lambda = 1\ \mu l. = 10^{-6}$ liter) of each sample has been applied. The spots may be labeled in *pencil* for identification purposes. Slits should be cut in the paper to insure equal vapor pressure and to prevent the mixing of spotted material during development (Fig. 4.2D). Insert the wick in the center hole and place the paper on the bottom pie plate so that the wick extends into a small vessel of solvent placed in the center of the plate. (Remember to mix the solvent thoroughly just prior to use). Put the top plate in place and allow the paper to develop for three to four hours, that is, so that the solvent nearly reaches the edge of the paper. At this point remove the paper and *quickly* mark the solvent front with

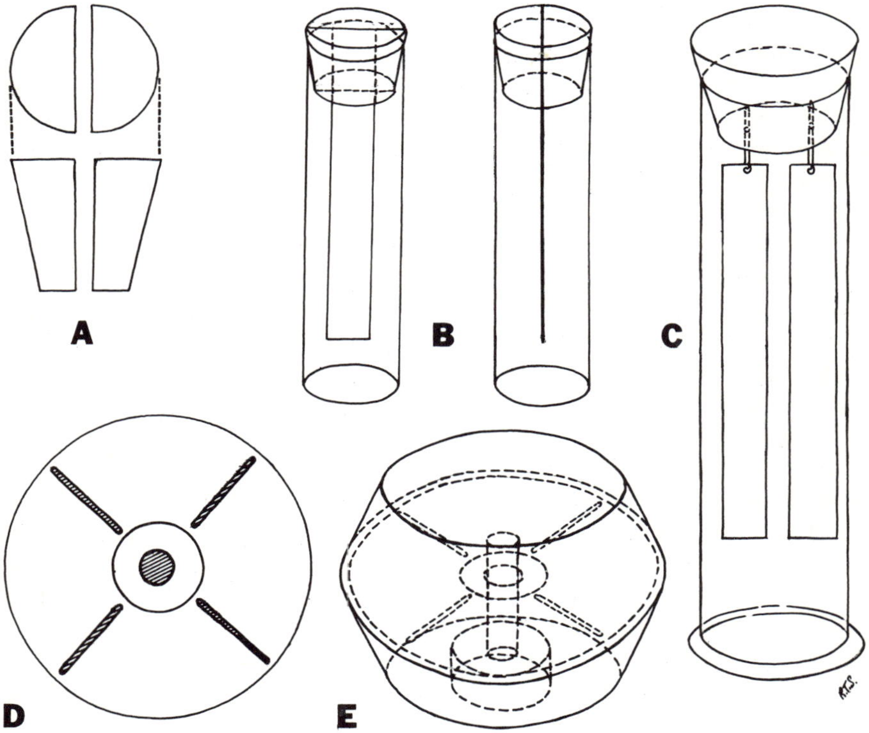

Figure 4.2. Apparatus for chromatography. (A) bisected stopper; (B) two views of cylinder with bisected stopper in place holding paper strip; (C) cylinder with paper suspended from bent pins; (D) circular paper with center hole for wick, concentric pencilled circle and four radial slits; (E) pie plate apparatus assembled.

a pencil. When the paper is dry, spray with ninhydrin. Mild heat may aid color development.

The ratio of the distance traveled by the substance to the distance traveled by the solvent is known as the R_f **(ratio of fronts).** Calculate the R_f of ninhydrin-positive spots in hemolymph and compare these to the R_f's of the known amino acids which you have chromatographed. Can you identify any of the unknown spots? Do the free amino acids in the hemolymph of insects of the same species vary? Are there significant differences among species? Where possible, compare your results to those obtained from the more gross chromatographic investigations outlined in Chapter Two. The present exercise should suggest additional modifications and refinements for the previous studies of insect systematics.

C. RESPIRATION

The measurement of whole insect respiration has long been an important physiological problem, since reliable values of oxygen consumption and/or carbon dioxide production underlie a great many related problems. The majority of respirometry is accomplished using various manometric devices (Dixon, 1951), the standard being the Warburg apparatus. In most studies involving closed systems the insect is kept under conditions of gradually reduced oxygen tensions. This changing oxygen environment, together with controlling (or at least determining) the activity of the experimental animal, are the most troublesome problems in whole animal respirometry. As Krogh (1916; 1941) and a host of others have demonstrated, respiration is not a constant but varies significantly with conditions within and outside the organism under study.

Insect physiologists have typically utilized values of volume or weight of oxygen consumed per gram of insect per unit of time (hr.) to express general metabolic activity of the organism or its tissues. This has been accomplished under various physical and chemical conditions of the environment and for various life stages and activity states of the insect species being tested. The ratio of carbon dioxide produced to oxygen consumed (CO_2/O_2) when both are expressed in the same units (*e.g.*, cc./g./hr.), which is known as the **respiratory quotient** (RQ), has been employed to determine the class of organic food being metabolized by an insect.

Good reviews of respirometry (Krogh, 1916; Tobias, 1943; Dixon, 1951) and insect respiration (Edwards, 1953; Wigglesworth, 1956; Nat. Acad. Sci., 1958; Buck, 1962; Patton, 1963) are available.

The following exercises are designed to introduce the entomology student to simple respirometry techniques and to stress the relationship between the experimental environment and the physiological response. The procedures for measuring rates of oxygen consumption by a terrestrial insect will be utilized in a later section dealing with physiological rates in more general terms of temperature coefficients.

Data on oxygen consumption and carbon dioxide production by aquatic insects introduce some basic procedures of water analysis and allow the computation of RQ for a series of temperatures. As will be noted, such RQ values provide important data on nutrition and energetics.

Although no exercises are included on the respiration of specific insect tissues as measured with the Warburg apparatus, such experiments can be added for advanced groups. References on methodology (Dixon, 1951; Umbreit, Burris, and Stauffer, 1957) and typical results (Bliss and Skinner, 1963) should be consulted. The technique is especially valuable in studying the effects of particular metabolic inhibitions, enzymes or hormones on insect tissues.

1. OXYGEN CONSUMPTION BY TERRESTRIAL INSECTS

MATERIALS. For small insects with low rates of oxygen consumption the Warburg respirometer has been used successfully, although it is expensive and requires considerable operating skill in order to obtain reproducible results. Other highly refined methods such as the Scholander and Edwards (1942) microrespirometer have been figured by Dixon (1951) and Prosser *et al.* (1952: Fig. 44). Both simple and complex respirometers are shown in Welsh and Smith (1960); see also the review by Scholander (1952). An extremely complex respirometer, which avoids the problem of oxygen depletion, has been developed by Winteringham (1959).

The basic parts of the respirometer are two wide-mouth metric laboratory bottles fitted with rubber stoppers (Fig. 4.3). The bottles are connected by a piece of capillary glass tubing (1) having an evenly bent curve at the center with an end fitting into each stopper; the center portion is filled with manometer fluid. A white card should be taped to the curved manometer and marked to indicate the fluid level at each end when pressure is equal. A stopcock is fitted into a hole of each stopper (2) and all connections, including the stopper-bottle interfaces, are sealed with stopcock grease. The needle of a 1 to 10 cc. hypodermic syringe (3) is inserted through the stopper into the chamber that contains the experimental insects. A third hole should be bored in this stopper to accommodate a centigrade thermometer (4). The volume of the bottles and the syringe can be adjusted to provide the maximum accuracy possible for any given insect species.

The volume of the insect or insects to be tested should be relatively large in comparison to the volume of the bottles—25 to 150 ml. capacity bottles are usually the most convenient. The second bottle is a thermobarometer which compensates for pressure or temperature changes in the system.

A shell vial containing 20% KOH (5) is taped to the inside of the bottle serving as the animal chamber. A piece of filter paper about one and one-half times the length of the vial and folded in pleated fashion is placed in the vial. The amount of KOH used should be just enough to

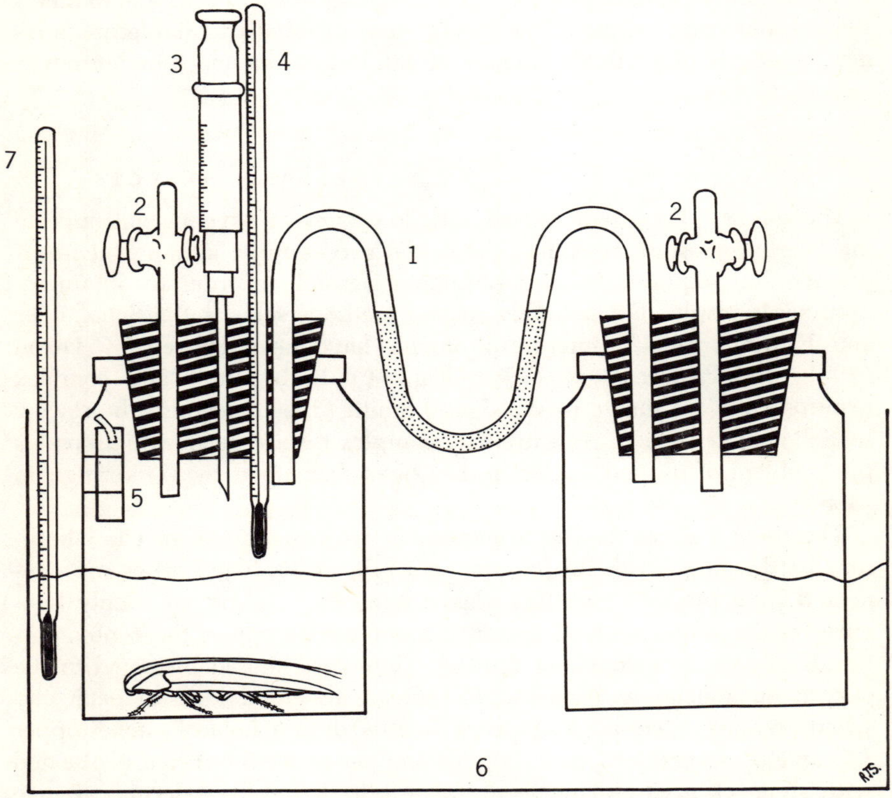

Figure 4.3. Two-chambered respirometer for terrestrial insects. 1, manometer tube; 2, stopcock; 3, hypodermic syringe; 4, thermometer; 5, shell vial containing 30% KOH; 6, water bath; 7, water bath thermometer.

saturate the paper. This arrangement serves to absorb CO_2 produced within the respirometer. The respirometer vessels must be immersed in a water bath (6) fitted with a thermometer (7).

The insects utilized will depend, of course, on availability. Cockroaches (*Blaberus*) permit large insects to be compared with a smaller form (*e.g.*, *Periplaneta*) in the same suborder. Small insects such as grain beetles (*Tenebrio* and *Tribolium*) and fruit flies (*Drosophila*) can be utilized to test the limitations of measurement with such a respirometer. Larvae, pupae and adults of *Tenebrio molitor* provide convenient material for comparing respiration in different life stages.

A balance with sensitivity that will permit weighing insects to the nearest 0.01 g.; a suitable container such as an enamel tray or an aquarium for the water bath; crushed ice; hot tap water; anesthetic (CO_2 or hypothermia); timer; black electrician's tape for darkening respirometer chambers as desired.

METHODS. At least three temperature regimes should be used for each life stage of each species tested; it is convenient to use room temperature, a warmer temperature with a hot water bath and a cooler temperature with a bath of crushed ice in water. Place the respirometer in the water bath and allow ten to twenty minutes for the system to come to equilibrium. When the thermometers in the respiration chamber and the water bath read the same, the experiment should begin.

Fill the syringe with oxygen. To do so, attach a piece of plastic tubing to an oxygen tank outlet. Turn on the gas and force the air out of the tube, clamp the distal end of the tube and then the end near the tank. Insert the syringe needle through the tube wall and fill.

One or more individuals of the insect to be tested are introduced into the respiration chamber and the stopper replaced, keeping both stopcocks open. The syringe is inserted part way into the chamber and the plunger set exactly at a graduation. When ready to start measurement, close both stopcocks. Equalize the pressure in the two chambers as indicated by the manometer fluid, which should be centered.

Start timing the experiment when the fluid has been centered, the system completely closed and the initial syringe setting and temperature readings taken. As the insect uses oxygen, pressure is reduced in the animal chamber because the CO_2 of respiration is absorbed by the KOH —accordingly, the fluid rises on the side adjacent to the animal chamber. Every five to fifteen minutes restore the original level of the manometer fluid by adjusting the syringe plunger.

The volume of oxygen added during the time period of the experiment (t) is assumed to be equal to the oxygen consumed (cc. O_2) by the enclosed insect. Following the experiment, the insect is anesthetized and weighed in grams. The consumption of oxygen per gram of insect per hour can be computed for the particular experimental temperature using the formula

$$[(\text{cc. } O_2 \text{ utilized})\ (60/t \text{ min.})]/\text{insect wt. in g.} = \text{cc. } O_2/\text{g./hr.}$$

A control series, with no insects in the chamber, should be run for time intervals comparable to experimental periods to evaluate the reliability of the system. It is important that the animal respiration chamber be thoroughly washed after each experiment and fresh KOH added to the shell vial.

After a number of trials have been run for a particular life stage of a given species at each of three temperatures, the results can be plotted with cc. O_2/g./hr. against temperature, giving a graph of the kind shown in Figure 4.5.

Similar graphs can be derived for other life stages of the same species, for each sex and for different species. It is also interesting to compare the results of experiments conducted in the light with those run in the dark. In this way curves for active and inactive individuals can usually be obtained. The activity of experimental animals is considered further in the next section.

Insects such as *Ranatra fusca* de Beauvois, the water scorpion, which are aquatic but actually utilize atmospheric oxygen, can be tested in the respirometer by partially filling the animal chamber with water; for treatment of water in such an experiment, see the following exercise.

Some insects are known to have a discontinuous respiration characterized by bursts of CO_2 release (Buck and Keister, 1955); experiments designed to run for longer periods of time reduce the probability of obtaining erroneous results.

2. OXYGEN CONSUMPTION BY AQUATIC INSECTS

Respirometry with aquatic insects presents some special problems, although manometric devices can also be utilized in such experiments (Morris, 1963). Because there is often a considerable lag in the establishment of oxygen equilibrium across the water–air interface in experimental vessels, it is usually more satisfactory to measure the dissolved oxygen concentration directly. This can be accomplished by using specialized and expensive oxygen electrode equipment such as the Beckmen Model 777 polarographic oxygen sensor and laboratory analyzer (see Appendix), or by the relatively simple but quite accurate technique described below.

Materials. A suitable respiration chamber is shown in Figure 4.4. It consists of a wide-mouth aspirator bottle (50–250 ml.) with tubulature to which an 8 cm. piece of vacuum pressure tubing (1) is fitted; the end of the tubing is closed with a pinch clamp. Diffusion of oxygen into or out of the water is prevented with a 2 to 4 cm. layer of heavy (S.A.E. 20 or 30) motor oil (2). A two-hole rubber stopper is fitted into the mouth of the bottle; one hole takes a centigrade thermometer (3), the

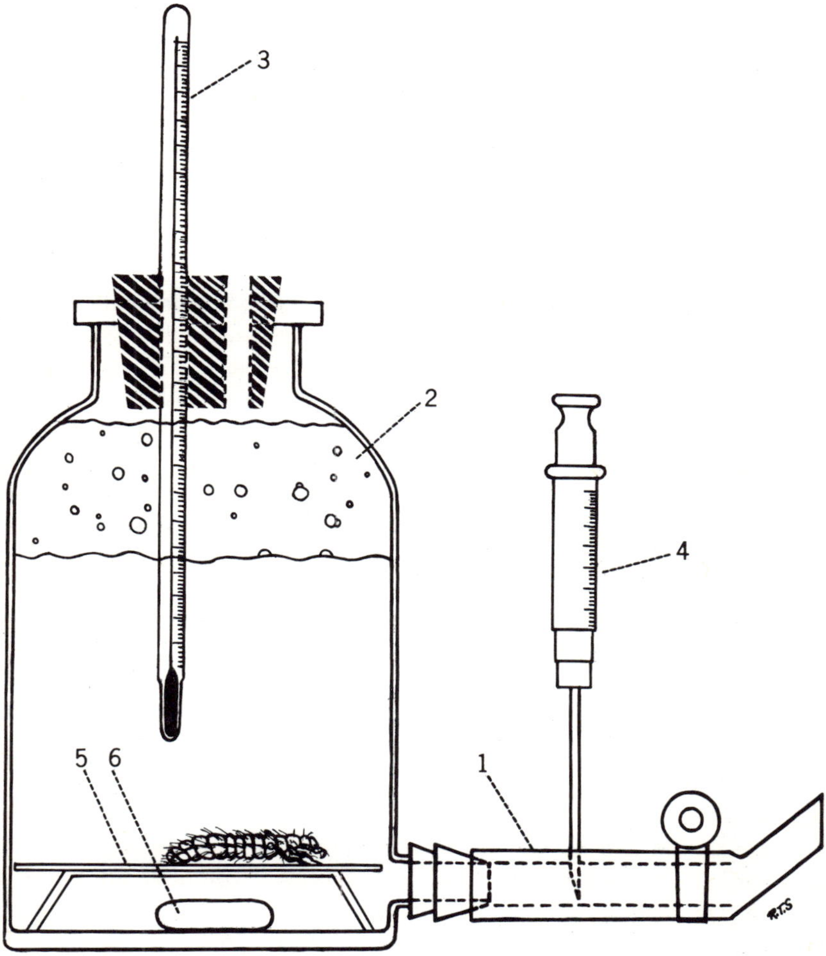

Figure 4.4. Aspirator bottle respirometer for use with aquatic insects. 1, thick walled rubber tubing; 2, motor oil; 3, thermometer; 4, 30 ml. hypodermic syringe; 5, aluminum screen and support; 6, Teflon coated stirring bar.

other is left open. Samples are withdrawn from the thick-walled tubing with a 30 ml. (cc.) hypodermic syringe (4). An aluminum screen support (5) for the experimental animals, under which is placed a teflon coated stirring bar (6), is optional.

Great care must be taken to insure that only insect respiration is measured in the experimental chamber. Therefore the bottles must be sterile at the outset, the aquatic insects to be tested must be washed thoroughly in distilled water and special precautions must be taken concerning the culture water to be utilized. Distilled water can be used with satisfactory results (Eriksen, 1961). If tap water is used it must be conditioned (Appendix). Natural stream, pond or lake water will suffice if it is filtered first using a 0.45 μ pore size, Millipore filter (see Appendix and Chapter Six). Indeed, if the equipment is available, it is desirable to filter just prior to experimentation any water to be used. Whatever the experimental water employed, it should be thoroughly aerated prior to use, except in the O_2 consumption versus O_2 concentration exercise below. An aquarium pump and air stone system is best.

Analysis of dissolved oxygen is achieved by a minor modification (Lund, 1921; van Dam, 1935; Krogh, 1935; Fox and Wingfield, 1938; Eriksen, 1961; Sessions, 1961) of the Winkler method (Welsh, 1948; Amer. Pub. Health Assoc., 1960). It has been established (Allee and Oesting, 1934) that the Winkler method is sufficiently accurate for respiration studies on aquatic organisms if it is carefully handled. The method depends upon a series of chemical reactions which liberate free iodine in amounts chemically equivalent to that of the dissolved oxygen that was present in a water sample. The amount of this free iodine is determined by titration with sodium thiosulfate, using an iodometric indicator dye.

Reagents needed (Amer. Publ. Health. Assoc., 1960): Manganous sulfate solution, $MnSO_4$ (dissolve 480 g. $MnSO_4 \cdot 4H_2O$, 400 g. $MnSO_4 \cdot 2H_2O$ or 364 g. $MnSO_4 \cdot H_2O$ in distilled water, filter and dilute to 1 liter; the specific gravity of the solution should be 1.270 at 20° C.). Alkali-iodide, NaOH-NaI (dissolve 500 g. NaOH, or 700 g. KOH, and 135 NaI, or 150 g. KI, in distilled water and dilute to 1 liter). Concentrated orthophosphoric acid, H_3PO_4. Sodium thiosulfate, $Na_2S_2O_3$, 0.001*N;* this normality of thiosulfate solution can be conveniently prepared from 0.1*N* or 0.01*N* stock solutions; the stock solutions should be obtained from a reputable chemical supplier (Appendix) to insure best accuracy. If stock solutions are prepared, see Amer. Publ. Health Assoc. (1960). A 0.001*N* sodium thiosulfate solution can be prepared according to the following formula:

$$\frac{(1000)\ (\text{desired normality})}{\text{normality of stock solution}} = \text{ml. of stock solution required to make 1 liter of the desired normality}$$

If a $0.01N$ stock solution is used, 100 ml. are required to make 1 liter of $0.01N$ solution for titrations.

$$\frac{(1000)\,(0.001)}{0.01} = 100$$

Thyodene is a stable, soluble, dry indicator for iodometric titrations; a starch solution can be utilized as an alternate. (Prepare 5–6 g. soluble starch in a beaker; add a little distilled water and mix. Pour this emulsion into 1 liter of boiling water; allow to boil 3–5 minutes and let it settle overnight; use the supernatant.)

A 30 ml. capacity Leur-Lok type glass hypodermic syringe, with a stainless steel needle (Fig. 4.4) is required for removing water samples and mixing the reagents in chemical analysis. In addition, a 5 or 10 ml. capacity microburet, several 10 to 20 ml. flasks or beakers and a 500 ml. graduated cylinder are needed. A 10 ml. volumetric pipet is optional.

Any immature insect utilizing dissolved oxygen is suitable. The species, such as some hemipterans, some coleopterans and a few dipteran species, actually utilizing atmospheric oxygen should be tested in the respirometer described in the previous section. Since a number of other exercises in this book require trichopteran larvae, they are convenient subjects for this exercise. It is quite instructive to test species from both standing (**lentic**) and running (**lotic**) water habitats in order to compare oxygen requirements of species that naturally occur in environments differing with respect to dissolved oxygen content.

Water baths, balance, timer, anesthetics and electrician's tape (see previous exercise).

METHODS. Select the insects, rinse them thoroughly in distilled water and allow at least an hour for them to acclimate to the experimental water temperature employed. This can be accomplished by placing the animals in a vessel containing distilled water maintained in the same water bath containing the respirometer bottle. If a magnetic stirrer is available, the stirring bar and aluminum screen support should be placed in the aspiration bottle. The experiment is initiated by adding exactly 220 ml. of the appropriate water to the bottle (be sure the pinch clamp is tight), introducing the experimental animal(s), and floating a 2 to 4 cm. layer of heavy motor oil at the top of the bottle; insert the two-hole stopper fitted with a thermometer. Remove a 20 ml. sample of water through the rubber tubing with the hypodermic syringe. Record the water temperature and the initial time.

The 20 ml. water sample should be analyzed as follows:

1. Draw approximately ¼ ml. (0.25) each of the $MnSO_4$ and alkali-

iodide reagents into the syringe and shake. Since excess reactions are involved, the exact amount of reagent added is not critical—between 0.1 and 0.5 ml. is acceptable.

2. Draw approximately ¼ ml. of concentrated H_3PO_4 into the syringe and shake again; if some precipitate still remains, add a little more acid.

3. Expel the solution from the syringe into a 30 ml. flask and draw two 10 ml. subsamples with a volumetric pipet or measure the 10 ml. amounts directly from the syringe. Place each in a 10 to 20 ml. flask or beaker for titration and add a small amount of thyodene to produce a dark blue color.

4. Titrate with 0.001*N* sodium thiosulfate, using the microburet, until the solution flashes to clear (disregard reversal of the end point which may occur if the solution is allowed to stand). Record the number of milliliters of thiosulfate used in titration (initial ml. thiosulfate). The two titrations should agree exactly.

Suitable experimental periods range from about fifteen minutes to two hours. The culture water should be stirred briefly several times during the experiment and just prior to removal of a second water sample at the conclusion of the experiment. If a magnetic stirrer is not available, the bottles should be gently swirled during the experiment and just prior to removal of water samples.

Record the time and water temperature when the final 20 ml. water sample is withdrawn. Remove the animal(s) from the chamber, blot them dry and weigh them to the nearest 0.01 g.; it may be necessary to anesthetize the insects, as described in the previous section. If case-bearing Trichoptera are used, remove their cases prior to weighing.

The change in oxygen concentration in the experimental bottle expressed in terms of milliliters (cc.) of sodium thiosulfate used in titration is:

$$(\text{initial ml. thiosulfate}) - (\text{final ml. thiosulfate}) = O_2 \text{ used as ml. thiosulfate}$$

This volume of thiosulfate is multiplied by a correction factor (described below) to obtain cubic centimeters O_2 consumed by the insect(s) during the experimental time interval.

The milliliters of sodium thiosulfate utilized in a titration is exactly equivalent to the milligrams per liter (1 mg./liter = parts per million or ppm.) of dissolved oxygen if 0.025*N* thiosulfate is used to titrate a 200 ml. sample. When 0.001*N* thiosulfate is used, a correction of 0.04 (0.001/0.025) is necessary to obtain results in mg./liter. If a 10 ml. sample is titrated, a further correction of 20 is necessary (200/10). Thus, the complete correction can be accomplished as follows:

$$(\text{ml. thiosulfate used})\ (0.8) = \text{mg./liter } O_2$$

This value can be converted to cubic centimeters per liter:

$$\text{cc./liter} = (\text{mg./liter})\ (0.698)$$

In order to correct the per liter values to the absolute amounts in 200 ml. of water in the experimental chamber, a correction factor of 0.2 (200/1000) is necessary. The entire correction can be expressed as:

$$\text{cc. } O_2/200 \text{ ml.} = (\text{ml. thiosulfate used})\ (0.16)\ (0.698) = (\text{ml. thiosulfate})\ (0.1117)$$

The value obtained for cubic centimeters O_2 consumed (O_2 used as ml. thiosulfate × 0.117), can be converted to a per gram per hour basis using the formula given in the previous exercise.

It is important that a control bottle identical to the experimental ones, but containing no animals, be tested in each experimental run. Most workers conduct the experiments in the dark (Eriksen, 1963a, 1963b), which can be achieved by covering the bottles with electrician's tape. It is interesting to compare series tested in the light with those conducted in the dark. Most often the animals are inactive in the dark, thus yielding a "resting" respiration rate. Note too that respiration in young nymphs or larvae can be compared to mature individuals on a per gram basis.

Figure 4.5 shows a typical plot of respiration as cubic centimeters O_2 consumed per gram of insect per hour, against temperature for two species of limnephilid caddis fly larvae. *Pycnopsyche lepida* (upper curve) inhabits regions of streams where water is faster running than regions inhabited by *P. guttifer* (lower curve), as shown by Cummins (1961, 1964).

At a given temperature the respiratory rate of *P. lepida* is higher than that of *P. guttifer*. Although *P. lepida* may be slightly smaller than *P. guttifer*, both have a mean wet weight close to 0.49. Thus, the differences in respiratory rate cannot be accounted for merely on the basis of the inverse relationship that exists between body size and oxygen consumption (Prosser and Brown, 1961). A higher rate of oxygen consumption by species of fresh-water invertebrates from rapidly flowing water as opposed to lower rates for species in slow flowing or standing water has been noted by a number of investigators (Fox *et al.*, 1933, 1935, 1937; Berg, 1952, 1953; Berg *et al.*, 1958; Berg and Ockelman, 1959; Eriksen, 1963a, 1963b).

A comparison between oxygen consumption and oxygen concentration can be made by adjusting the dissolved oxygen level in the experimental water. The water can be deoxygenated either by boiling, followed by various periods of aeration, or by bubbling nitrogen gas through it for

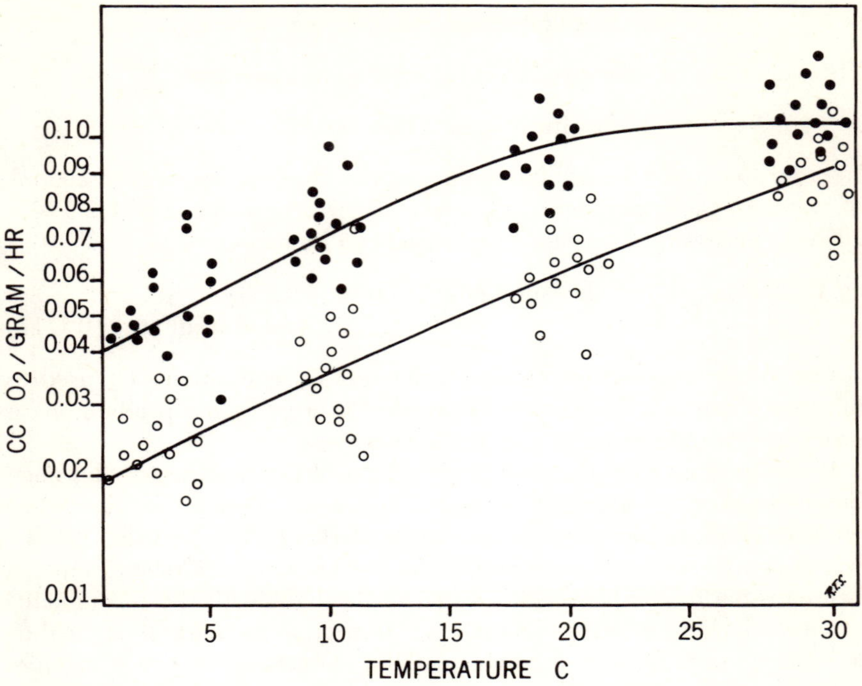

Figure 4.5. Comparison of respiration in terminal instar larvae of the caddis flies *Pycnopsyche lepida* (closed circles) and *P. guttifer* (open circles) as affected by temperature, drawn on semilog plot. (Redrawn from Sessions, 1961, by permission)

varying time intervals. The dissolved oxygen content is checked by the Winkler method at the outset of the experiment, as described above, and all experiments are conducted at the same temperature. A useful nomogram showing oxygen saturation values at different temperatures is given in Welsh (1948).

Figure 4.6 shows curves of oxygen consumption against oxygen concentration in the experimental water at 12° C. for the same two species of limnephilids shown in the previous figure. It is apparent that *P. guttifer* (lower curve) is able to maintain a rather constant respiratory rate over a wider range of oxygen concentration than is *P. lepida.* Since *P. lepida* lives in habitats where the water is usually better aerated, it is reasonable to expect that this species would be less tolerant of decreases in environmental oxygen.

The importance of the substrate upon which aquatic insect respiration is tested has been clearly shown by Eriksen (1961, 1963a, 1963b). He

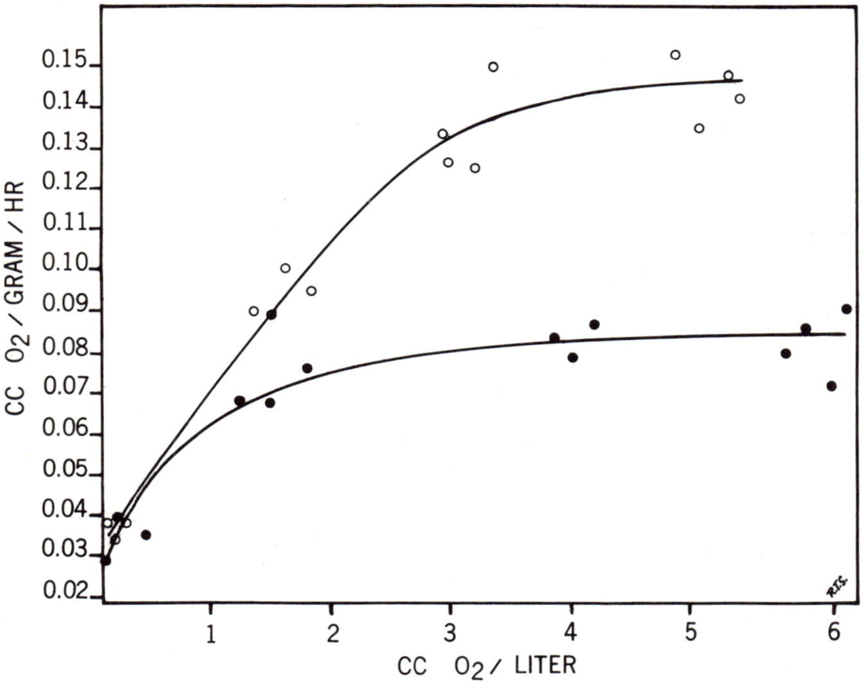

Figure 4.6. Oxygen consumption by *Pycnopsyche lepida* (open circles) and *P. guttifer* (closed circles) relative to dissolved oxygen concentration at 12° C. (Redrawn from Sessions, 1961, by permission)

found that the rates of oxygen consumption by two species of burrowing mayflies in "blank bottles"—bottles in which the animal was not provided with a substrate—were significantly different from rates of consumption when the nymphs were provided with mineral substrates. He demonstrated an optimal particle size range for each and showed that gill structure and function are most important in regulating oxygen consumption.

Various substrate materials can be tested conveniently in the respirometer described above (Fig. 4.4), but all such materials must be sterilized (autoclaved) prior to use in the experimental bottles.

3. CARBON DIOXIDE PRODUCTION BY AQUATIC INSECTS AND DETERMINATION OF RQ

The measurement of carbon dioxide production is necessary in order to calculate the **respiratory quotient** (RQ), defined as the ratio of CO_2 produced to O_2 consumed. As indicated previously, the RQ value (CO_2/O_2) for a given species under particular conditions can be employed to estimate the general kind of organic material being metabolized. Carbohydrate metabolism produces an RQ of 1.00, fats 0.71 and proteins 0.79.

In general, herbivores tend to have RQ values close to 1.0 and predators have values nearer 0.79. Ecologists have used RQ values to estimate the cost of respiration, in energy expenditure, to various organisms (Englemann, 1961). The energy expenditure of respiration is termed the **maintenance cost** of an organism or a population. If the rate of oxygen consumption is known, together with an RQ value, the **energy equivalent** (calories) for respiration can be calculated using the data of Brody (1945).

MATERIALS. Respiration chamber and insects as described in the previous section; pH meter (Beckman Zeromatic); 0.01*N* NaOH and 0.01*N* H_2SO_4, which can be purchased in these normalities or prepared from stock solutions of known normality, as described in the previous section; glassware as described in previous section; experimental water: distilled water (filtered if possible) is satisfactory but the water employed *must* have a minimum buffering capacity, since the measurements depend upon a pH shift.

METHODS. This experiment follows the same procedure given above for measuring oxygen consumption, except that samples withdrawn with the hypodermic syringe are placed in small beakers for pH analysis. It may prove convenient to withdraw such samples for pH analysis in conjunction with experiments on oxygen consumption. However, the lack of a buffered water system is not critical in oxygen measurements and in fact it may be desirable to use a buffered system (Fox, *et al.*, 1935).

Prior to the experiments, a curve of the buffering capacity of the water system, relative to the addition of acid and base, must be constructed. A hypothetical curve of the type required is shown in Figure 4.7. The changes in pH compared to the initial pH reading, resulting from the addition of each ml. of 0.01 H_2SO_4, are plotted. The pH shift caused by 1 ml. of 0.01 H_2SO_4 is chemically equivalent to the shift caused by the addition of 10 micromoles (μm.) of CO_2. Conversely, the shift due to each 1 ml. of 0.01 NaOH is equivalent to the removal of 10 μm CO_2. This latter relationship has been used in photosynthesis studies (Verduin, 1951).

Once a curve has been plotted for the experimental water, any change in pH can be related to the equivalent titration change obtained with the acid and therefore to a CO_2 change.

The amount of CO_2 produced during an experiment can be converted to cubic centimeters CO_2 per liter as follows:

$$(\mu\text{m. } CO_2)\ (44)\ (10^{-3}) = \mu\text{g./liter}\ (0.698) = \text{cc. } CO_2/\text{liter}$$

This value is then corrected for the experimental time duration and for animal weight, as described in the previous exercise, in order to arrive at values expressed on a per gram per hour basis.

In most cases, it is desirable to adjust the pH of the experimental water slightly by the addition of several milliliters of 0.01*N* NaOH. This tends to center the pH shifts about the original pH of the culture or experimental water, to which the organisms are presumably already acclimated. Although CO_2 production is often measured by other methods, such as collecting the gas on a preweighed absorbent and weighing the CO_2 produced, or with C^{14} as a tracer, the technique described above is sufficient to obtain a reasonable estimate.

An RQ value can be computed for each species for which data have been gathered on both oxygen consumption and carbon dioxide produc-

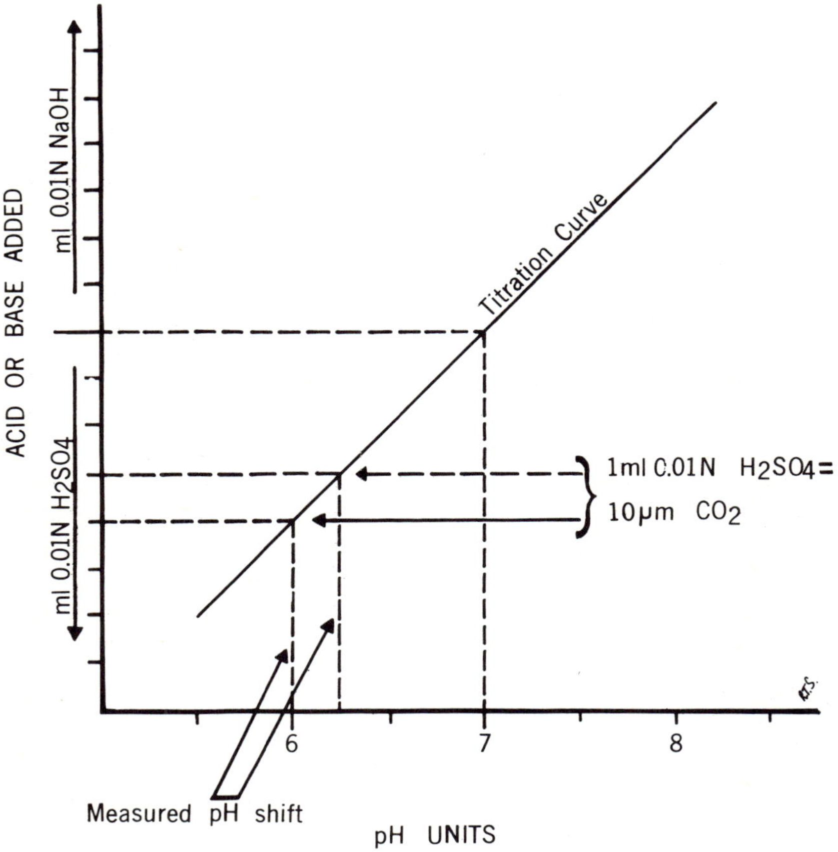

Figure 4.7. Hypothetical titration curve for the pH method of determining CO_2 production by aquatic insects.

tion at the same temperature. The calculated values should be compared to those given above for each of the classes of organic compounds. If data are available on the natural food habits of the species (see Chapter Six), the relative importance of carbohydrate fat and protein digestion with regard to the calculated RQ values can be considered.

D. THE TEMPERATURE COEFFICIENT Q_{10}

The effect of temperature on the biological processes of animals, especially poikilotherms, is an important consideration. Insects lend themselves particularly well to studies of these phenomena. The temperature effect can be quantified, using the following formula to express the factor by which a given reaction rate will be increased by a 10° C. rise in temperature (the Q_{10}):

$$Q_{10} = (K_1/K_2)\ (10/T_1 - T_2)$$

where T_1 refers to the higher temperature, T_2 to the lower temperature, K_1 to the rate of reaction at T_1 and K_2 to the rate at T_2. Note that the computation of Q_{10} by the above formula does not require an exact 10° C. temperature change, but may be used for any temperature interval.

Values for Q_{10} obtained with this formula will vary according to the nature of the reaction under consideration. Normal values of Q_{10} for ordinary chemical reactions range from 2 to 3, but photochemical reactions are generally lower. For a discussion of Q_{10} see Giese (1963: 199–203) or other physiology reference sources, since this reading will be of value in the interpretation of results.

In the experiment which follows, Q_{10} values should be calculated for all trials and a graph prepared plotting reaction rates against temperature.

MATERIALS. Experimental insects used in previous experiments on respiration (Section C), both terrestrial and aquatic; respirometers (Figs. 4.3, 4.4); Pringle's solution (Appendix), and some to which 0.01% indigo carmine has been added; 50% ethyl alcohol; stopcock or vacuum grease; dissecting dishes and pins; watchmaker's forceps (no. 5 size); microdissection scissors; dissecting microscopes; funnels and filter paper; spectrophotometer and cuvettes; water baths; thermometers; ice; heat source for warming.

METHODS. The three following exercises examine the effects of temperature on oxygen consumption of terrestrial and of aquatic insects respectively, on insect heart rate and on dye transport by the malpighian tubules.

1. EFFECT OF TEMPERATURE ON OXYGEN CONSUMPTION (QO_2)

The cockroach (or other terrestrial insect) is weighed and placed in the respirometer as shown in Figure 4.3; the procedure is exactly as outlined in the section on oxygen consumption by terrestrial insects (Section C 1). In fact, new measurements need not be made if data are already available from the previous exercises.

As before, oxygen consumption (QO_2) is expressed as cubic centimeters (ml.) O_2 per gram of body weight of animal per hour. Values for at least two different temperatures are required. Using the formula, calculate Q_{10} for oxygen consumption.

Caddis fly larvae (or other aquatic insects) are weighed and placed in the respirometer shown in Figure 4.4; follow the steps outlined in the section on oxygen consumption by aquatic insects (Section C 2). Again, previously obtained data may be used to calculate the Q_{10}.

Q_{10} values for oxygen consumption by three species of caddis fly larvae over two ten-degree temperature ranges are given in Table 4.1. Oxygen consumption by *Pycnopsyche lepida* and *P. guttifer* has been discussed previously (Section C 2). Note that in the range of 20° to 30° C., *P. guttifer* maintains the best Q_{10}; in its natural habitat this species encounters the widest fluctuations in oxygen and temperature. *Molanna uniophila*, an inhabitant of wave-swept lake shores, lies intermediate to *P. guttifer* from slow-water areas of streams and *P. lepida* from stream situations with considerable current velocity.

Table 4.1 Q_{10} values for oxygen consumption over two temperature ranges for three species of trichopteran larvae in terminal instars [a]

Species	Q_{10} Values	
	10–20° C.	20–30° C.
Pycnopsyche lepida	1.5	1.2
Molanna uniophila	2.0	1.3
Pycnopsyche guttifer	2.1	1.6

[a] Data from Sessions (1961).

2. EFFECT OF TEMPERATURE ON INSECT HEART RATE

Aesthetize a cockroach with CO_2 or by chilling, and decapitate; make a lateral incision around the entire integument. Pin the animal on a wax surface dorsal side down; with forceps pick up the posterior sternites.

Peel the ventral surface forward, cutting away any clinging tissues, and discard. Locate the esophagus along the midline of the thorax, lift it and pull posteriorly with the forceps, exposing the long tubular heart. Cut the gut as near as possible to its posterior attachment, taking care not to spill fecal material onto the heart. Save the gut in a vessel of Pringle's solution for the following part of the experiment. Cover the beating heart with Pringle's solution at room temperature; note that the dissection provides a natural receptable for the saline. A half-hour should be allowed after the operation before readings are taken. The frequency of heartbeats should be determined under a dissecting microscope. Count the contractions of the heart at room temperature, then record the frequencies of heartbeats after addition of saline at various temperatures between 0° and 30° C.

The normal frequency of heartbeats of *Periplaneta americana* heart preparations may vary from 35 to 120 beats per minute. This will vary greatly, especially if the preparation was damaged during dissection. To determine whether the alary muscles are intact, the preparation may be stained *in vitro* with congo red (Smith, 1964).

A diagram of the circulatory system of *Blaberus trapezoideus* Burmeister, a cockroach closely related to *B. giganteus* used in the morphology section, is given by Fox and Fox (1964: 150, Fig. 5.4). This drawing should be used by the student to find the necessary morphological "landmarks."

3. EFFECT OF TEMPERATURE ON DYE TRANSPORT BY THE MALPIGHIAN TUBULES

This experiment is adapted from Ralph (1962: III-15-III-16).

Locate the thin yellow malpighian tubules at the junction of the mid- and hindgut of the gut saved from the preceding part of the experiment. Cut the gut transversely just anterior and posterior of the tubule attachments; wash the segment containing the tubules in Pringle's solution to rid it of debris and gut contents. Pick away the small white fat bodies with fine forceps to minimize adsorption of the dye. Prepare the malpighian tubules of three animals in this way.

Transfer the tubule masses into three separate *clean* containers in which identical volumes of the Pringle's solution plus dye have been placed. Allow one container to stand at room temperature, one in a dish of finely crushed ice and the other in a 30° C. water bath. Gently swirl the fluid in each container every fifteen minutes; after one hour transfer the preparations to dishes of fresh Pringle's solution to rinse away any adsorbed dye. Remove each of the tubule masses to separate test tubes containing 5 ml. of 50% ethyl alcohol and shake the tube to elute both the indigo carmine and the yellow pigment of the tubules (riboflavin). Filter off the tubules and other debris, and pour the supernatants into three cuvettes.

Read on the spectrophotometer at 610 μ to determine the amount of dye excreted. A reading at 450 μ should also be taken for each sample, since the yellow riboflavin content has been found to be directly related to tubule mass in *young healthy adult* cockroaches (B. J. Wall and C. L. Ralph, unpublished data). If a ratio of the absorbency readings 610/450 μ is used, the results obtained among different animals may be more accurately compared. For best results, all males (preferably) or all females should be used in this study.

Graph the results of the effects of temperature on this experiment, using the temperature as the ordinate and the 610/450 μ spectrophotometer readings as the abscissa. Evaluate the effect of temperature on the rate of dye transport.

E. NEURAL PHYSIOLOGY

The first of the two experiments detailed below requires minimal equipment and is easily performed; many variations and refinements should suggest themselves to any class. The second experiment can be undertaken only in well-equipped laboratories.

1. METAMERIC INDEPENDENCE OF THE INSECT NERVOUS SYSTEM

Unlike the more functionally centralized vertebrates, each insect segment constitutes an independent unit fundamentally capable of autonomous reflex behavior. This behavior, known as **metameric independence,** can be demonstrated on a gross behavioral level, as in this exercise, or by such more refined methods as electrically recording the spontaneous activity of ganglia, as discussed in the next exercise. Both studies are designed to aid the student in understanding the relative importance of the insect "brain" and the segmental ganglia. Reflex behavior is examined within an individual segment, as well as among segments.

MATERIALS. Cockroaches (*Blaberus, Periplaneta* or other); cotton-tipped applicator sticks; scissors (small); CO_2 for anesthesia; bell jars or other chambers to confine intact insect for study (any deep vessel may be used if the inside is covered with a thin coat of petroleum jelly at the top); test substances, such as ammonia, acetone, any of the contact insecticides, apple juice, strong smelling food juices, sugar solutions, etc.; fine drawn glass pipets.

METHOD. The behavior and responses of a normal intact cockroach should first be observed. After noting the running behavior, touch the animal with your finger on the antennae, cerci and other body parts. Blow mild puffs of air through a pipet at specific structures. Do the tactile receptors seem to be localized? Next dip a cotton swab into a solution of each of the test substances and place them, one at a time, near the animal. Observe the reaction and roughly estimate distances necessary to obtain reactions.

After noting normal reactions, the cockroach should be anesthetized and decapitated. Repeat the above procedure, noting any differences in general running behavior (prodding the animal may be necessary) and responses to the various stimuli. Remove the cerci and retest tactile stimuli.

Take note of the independence of thoracic and abdominal activity and try to evaluate the importance of the brain in mediating the previously established behavioral responses.

After observing the behavior of the decapitated roach, cut thorax from the abdomen, and test these isolated parts for any responses. Do they exhibit autonomous activity?

Other experiments may be performed in a similar manner following the removal of body parts in various combinations. How would the responses of the cockroach compare with those of an insect, such as *Rhodnius*, having a greater degree of cephalization (Maddrell, 1963)? Test various species of insects as time permits and compare results.

2. SPONTANEOUS ELECTRICAL ACTIVITY OF INSECT GANGLIA

It has been known for some time that the nerve cord and the separate ganglia of the cockroach, as well as those of other arthropods, will exhibit spontaneous electrical discharge. Either the entire nerve cord or isolated segments or ganglia may be used for demonstrations. Many pharmacological studies on spontaneous discharge have been generated by the work of Roeder and Roeder (1939), which should be consulted.

Correlation of spontaneous neural activity with behavior has been discussed by Roeder (1955, 1963) and study of his presentations will give the student some insight into the significance of spontaneous electrical activity in the nervous system of the intact insect.

MATERIALS. Cockroaches, ideally *Blaberus;* trichopteran larvae may also be used if available and should be studied for comparison; silver electrodes; connecting wires; preamplifiers (DC); oscilloscopes; insect saline (see Appendix); gases (O_2, CO_2); ion solutions (Ca^{++}, Na^{+}, etc.);

solutions of various pH values; various drugs, solutions of 10^{-6}, 10^{-5}, 10^{-4} and 10^{-3} g./ml. of the following are recommended: nicotine (tobacco smoke may be used), eserine (physostigmine), pilocarpine, acetylcholine, tetraethyl pyrophosphate (TEPP), atropine, picrotoxin; 10% ethyl alcohol; nerve chambers, if the entire nerve is to be removed (a source for obtaining a nerve chamber is given in the Appendix).

METHODS. Pin a cockroach on a wax surface with its dorsal side up and cut completely around the lateral edge of the exoskeleton. After removing the dorsal sclerotized elements, dissect away the gut and the other viscera exposing the nerve cord, which will remain attached to the ventral surface along the midline. Figure 4.8 shows the general structure of the central nervous system of the cockroach.

The electrodes may be hooked around the nerve cord, which is raised slightly, or the cord may be removed and placed in a nerve chamber (see Appendix). Either method is satisfactory, although it may be easier to record from a single ganglion *in situ*. In both cases care must be taken to prevent desiccation of the preparation by periodically moistening the tissues with insect saline.

The electrodes are coupled to a preamplifier and the amplified signal is then fed into an oscilloscope for demonstration purposes. Make sure the entire system is grounded.

Observe the normal activity pattern, noting the frequency and amplitude of the components of the pattern. Does this pattern change, or is it constant? If possible, let one untreated preparation remain on the electrodes for the duration of the laboratory period, making periodic observations. This preparation will also provide a "normal activity" pattern for later comparisons. Such a record may be made from one or several ganglia. Diagram a "normal activity" pattern for comparison with other later patterns.

Administer a few mild puffs of oxygen and of carbon dioxide and note the effects with each. Diagram the patterns obtained for these and following treatments, if significantly different than the "normal."

Test solutions of various pH values, such as mild acids and bases, by adding a drop to the preparation. Is there any change? What is the effect of increasing or decreasing pH on the activity pattern?

Test solutions of Na^{+}, Ca^{++} or other ions in the same way as above and note any changes in electrical activity. Try to determine the specific cause.

Test first the most dilute concentrations (10^{-6}) of drugs and note their effects. Nicotine, eserine (physostigmine) and pilocarpine should cause an increase in the level of activity in 10^{-6} concentrations; at higher concentrations they may abolish all activity. Acetylcholine and picrotoxin will increase the discharge. If time permits, test tetraethyl pyrophosphate (TEPP) and atropine (parasympatholytic); the latter will prevent action of pilocarpine if administered first.

To terminate the experiment, add a few drops of 10% ethyl alcohol to the preparation and note the effects after a few minutes.

Modifications. The entire system may be modified for obtaining a permanent record by mounting a camera in front of the oscilloscope screen. The electrical activity may also be dramatized by feeding the amplified signal through an audioamplifier and sound speaker.

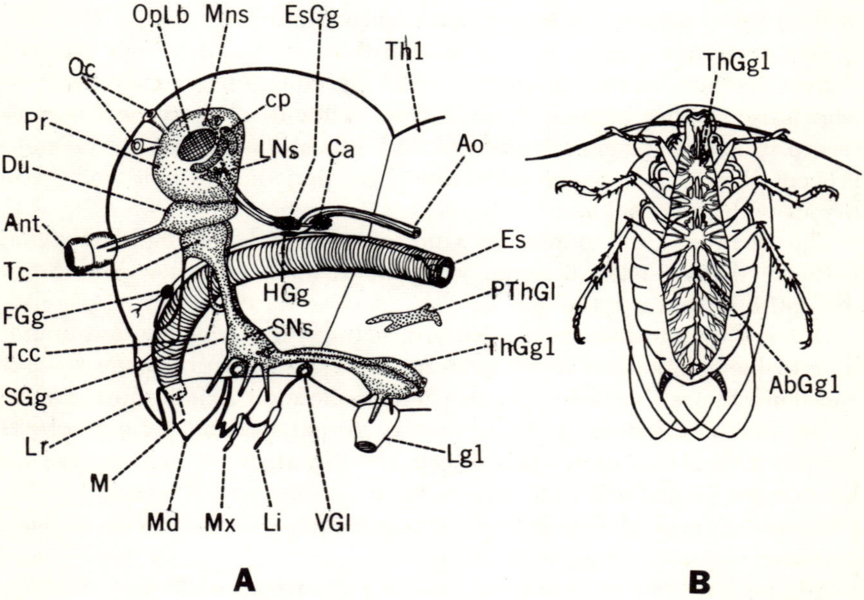

Figure 4.8. Nervous system of a cockroach. (A) parasaggital diagrammatic view of head and prothorax; (B) ventral view of dissection exposing the central nerve cord and ganglia. AbGg1, first abdominal ganglion; Ant, antenna; Ao, aorta; Ca, corpus allatus; cp, corpus pendunculatum; Du, deutocerebrum; Es, esophagus; EsGg, esophageal ganglion; FGg, frontal ganglion; HGg, hypocerebral ganglion; Lgl, prothoracic leg; Li, labium; LNs, lateral neurosecretory cells; Lr, labrum; M, mouth; Md, mandible; Mns, median neurosecretory cells; Mx, maxilla; Oc, ocelli; OpLb, optic lobe; Pr, protocerebrum; PThGl, prothoracic gland; SGg, subesophageal ganglionic mass; SNs, subesophageal neurosecretory cells; Tc, tritocerebrum; Tcc, tritocerebral connective; ThGg1, first thoracic ganglion; Th1, prothorax; VGl, ventral pharangeal gland. (A, redrawn from Jenkins, 1962, by permission of Pergamon Press)

F. HORMONAL CONTROL OF INSECT GROWTH AND DEVELOPMENT

In a recent review, Schneiderman and Gilbert (1964) have presented the current status of the endocrinology of insect growth and development, and one is impressed by the rate at which new advances are coming in this most exciting field. Another excellent review by the same authors (Gilbert and Schneiderman, 1961) lists over 225 references. Gilbert (1963) has reviewed (references through early 1961) the general field of knowledge relating to the endocrine control of molting and reproduction in invertebrates, particularly the crustaceans and insects (see also Bodenstein, 1953).

Figure 4.9, from Schneiderman and Gilbert (1964), summarizes the principal endocrine glands and the sites at which the hormones act for the cecropia silkworm moth *Hyalophora* (*Hyalophora*) *cecropia* (Linné). Neurosecretory cells in the *pars intercerebralis* of the brain secrete brain hormone (Bh) into the hemolymph by way of secretory complexes known as the *corpora cardiaca.* Bh stimulates the prothoracic glands (PrGl) to secrete prothoracic gland hormone (Ed) (**ecdyson**). Ecdyson in turn stimulates the growth of certain structures, particularly new cuticle, and initiates molting. Concurrent with the secretion of ecdyson, the *corpora allata* secrete juvenile hormone (Ne) (**neotenin**) into the hemolymph. Neotenin prevents metamorphosis to maturity; that is, during the larval stage, in the presence of ecdyson, the development of larval structures is promoted. As long as the balance of ecdyson and neotenin is maintained, the larva molts to another larval instar.

Termination of the larval period is marked by a cessation of neotenin secretion by the *corpora allata.* In the next molt ecdyson, in the presence of reduced titer of neotenin, initiates the development of pupal cuticle and, eventually, a complete transformation to pupal structures. At the last molt, no neotenin remains and ecdyson stimulates the epidermal cells to produce adult cuticle and eventually the other tissues are reorganized into adult structures.

Other functions ascribed to the *corpora allata* secretion in insects include direct or indirect control of egg maturation (except in Lepidoptera), accessory sex gland secretion, maintenance of larval and/or pupal diapause, mating and various aspects of general metabolism (Gilbert, 1963).

There is recent evidence (see review by Schneiderman and Gilbert, 1964) that ecdyson acts directly on the nucleus, as evidenced by increased RNA synthesis appearing as puffs in localized regions of the giant salivary chromosomes of dipterans such as *Tendipes* (= *Chironomus*). Increased

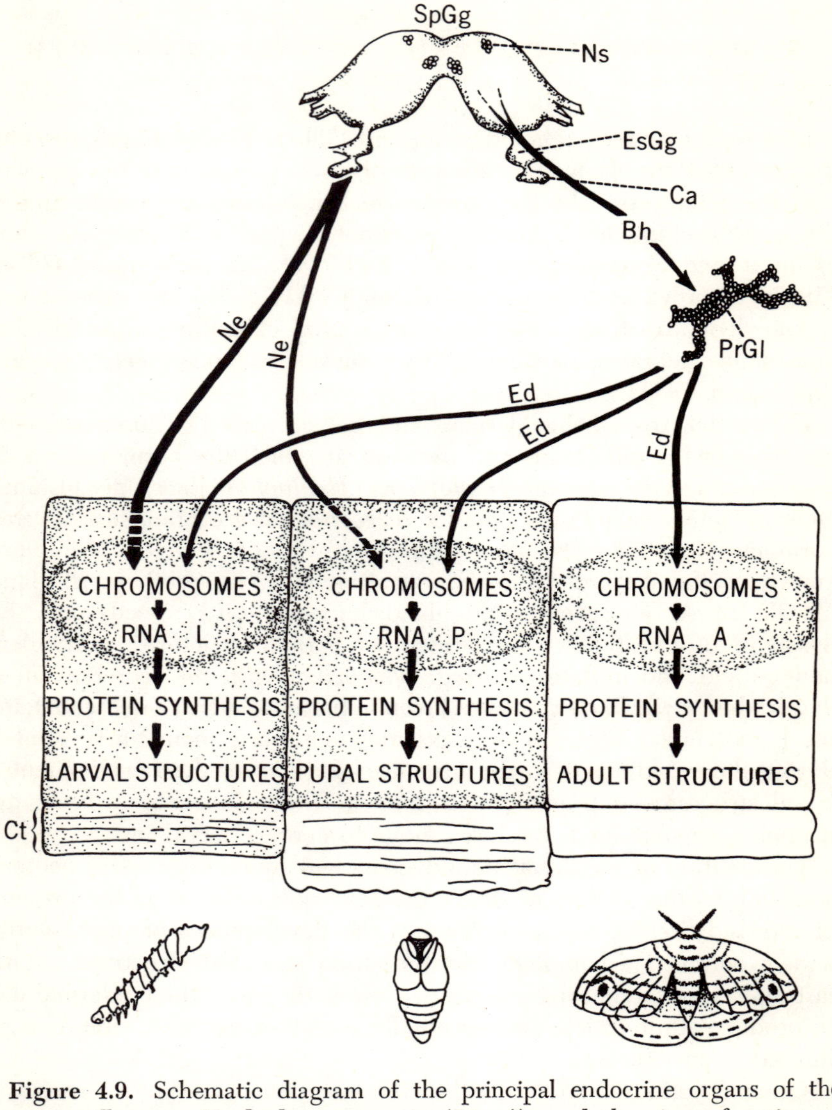

Figure 4.9. Schematic diagram of the principal endocrine organs of the cecropia silkworm, *Hyalophora cecropia* (Linné), and the sites of action of hormones in metamorphosis. Bh, brain hormone; Ca, corpus allatum; Ct, cuticle; Ed, ecdyson (prothoracic gland hormone); EsGg, esophageal ganglion (corpus cardiacum); Ne, neotenin (juvenile hormone); Ns, neurosecretory cells; PrGl, prothoracic gland; SpGg, supraesophageal ganglionic mass (brain). (Redrawn from Schneiderman and Gilbert, 1964, by permission of *Science*)

RNA synthesis leads to increased protein synthesis, which is at the base of new structural development. These relationships are shown schematically in Figure 4.9.

For a complete discussion of the basic patterns of insect endocrinology, as well as variations on the basic theme, the reader is directed to the reviews cited above.

It is interesting to note that an abnormally prolonged presence of neotenin, for example by artificial injections, can produce additional larval instars resulting in giants (Williams, 1958). Taxonomists troubled with the identification of immature stages have been following with interest certain experiments in which precocious adults have been produced by removal of the *corpora allata* (Wigglesworth, 1959). Once the exact structure of the steroid ecdyson, first crystalized by Butenandt and Karlson (1954), is known, massive doses of artificially prepared hormone may also hasten the advent of the development of species-definitive adult structures. In this regard, the recently discovered "brain hormone" activity of commercial preparations of cholesterol is of considerable interest (see Schneiderman and Gilbert, 1964).

1. EXTRACTION OF JUVENILE HORMONE

The hormone was first isolated by Williams (1956) from the abdomens of male cecropia moths, *Hyalophora* (*Hyalophora*) *cecropia* (Linné). Subsequent extractions have been made from the *corpora allata* of males of other giant silk moths (Saturniidae), particularly those of *Antheraea polyphemus* (Cramer), but *H. cecropia* still remains the best source. Since the original discovery, a great volume of work has accumulated concerning methods of extraction of the hormone, its effect on various insects and concerning other chemicals which bring about the same effects.

MATERIALS. Adult cecropia moths (or other species of saturniid moths, if cecropias are unavailable); the moths are best attracted in the manner described earlier in this chapter (Section A 1); ether; multimixer or Waring blender; beakers; filtering apparatus; water bath (optional for the Williams method); vacuum oven (optional for the Gilbert and Schneiderman method).

METHODS. The original extraction techniques and observations upon the nature of neotenin by Williams (1956: 213) may be summarized as follows: Blend one or more abdomens from adult male cecropias in a Waring blender at room temperature with ether, petroleum ether, ethanol or methanol. Of these solvents, ether is probably the best. Separate the solvent by filtration and evaporate it in a water bath. The hormone is now recovered as a residual, golden-yellow, nontoxic oil of low viscosity. The abdomen of each adult male cecropia usually yields about 0.2 ml. of oil. This oil is insoluble in water and may be washed repeatedly. After the initial extraction, the oil is soluble in acetone. The hormone is heat-stable.

Perhaps the method of choice for extraction of neotenin is that suggested by Gilbert and Schneiderman (1961). This procedure is as follows: Homogenize abdomens of adult male saturniids in ether in a multimixer for ten minutes at 16,000 rpm. Let the homogenate plus ether stand overnight with intermittent shaking. Filter and re-extract the residue with ether as above and refilter. Wash the filtrate three times with distilled water, discarding the wash water. This procedure will leave a golden-brown ether solution of the hormone. Evaporate away the ether solvent. Further evaporate the residual oil in a vacuum oven for one hour at 60° C.

The juvenile hormone, extracted by either method, may be used for the experiments following or for others as determined by the instructor.

2. WIGGLESWORTH METHOD OF BIOASSAY

Various bioassay methods have been utilized to determine the juvenile hormone activity of a cecropia extract. The procedure suggested by Wigglesworth (1958) employing pupae of the grain beetle *Tenebrio molitor* is presented here. The extracts from adult male cecropia moths prepared in the previous exercise should be assayed as follows:

MATERIALS. Cecropia neotenin extract; peanut or olive oil; fine dissecting needles (*minuten nadlen* in pin vices); 2 to 4 dram vials; gauze and rubber bands; pupae of *Tenebrio molitor;* several 1 ml. hypodermic syringes having a fairly coarse needle; 1 ml. beakers; small gummed labels.

METHODS. Dilute the neotenin extract prepared in the previous exercise with peanut oil or olive oil according to the following ratios of hormone to oil: 1:0, 1:1, 1:2, 1:4, 1:16, 1:64. Dilutions are made with 1 ml. hypodermic syringes on a drop basis, that is one drop of neotenin and four drops of oil are added to a 1 ml. beaker and mixed. Each diluted extract, or oil control, to be tested is drawn into a 1 ml. hypodermic syringe.

Isolate recently molted pupae (about 12 hours after molting) of *Tenebrio molitor*. Place a minute drop of one of the diluted neotenin extracts on each of three abdominal sternites along the left ventral side. Along the right ventral side place a minute drop of the oil used to dilute the hormone on each of three abdominal sternites. Puncture the integument through each of the droplets with a *minuten nadlen* held in a pin vice.

Use a separate needle for the hormone punctures and the control (oil) punctures and discard them after use.

Place each pupa, prepared as described, in a small vial. The end of the vial is covered with gauze held in place by a rubber band. Label each vial as to the neotenin dilution and control used, date and time of application, and time after molting to the pupal stage if known. These vials should be kept under the same conditions as the *Tenebrio* stock cultures.

When the treated *T. molitor* pupae molt to adults (6 to 8 days later at 25° to 30° C.) spots of pupal cuticle remain if a sufficient neotenin extract is used. These patches show up as white spots against the normal dark brown integument of the adult. The Wigglesworth (1958) method of scoring is useful in expressing the results. These scores are as follows:

large white area = 4
intermediate sized white area = 3
small white area = 2
brown scar with an extremely minute white point in it = 1
entire brown scar = 0

The controls should receive a score of zero. Each patch is scored separately and the three are averaged for each side of each experimental pupa. Average scores given by Wigglesworth for various dilutions are shown in Table 4.2.

Table 4.2 The effects of neotenin extracts of various dilutions (neotenin: oil) on *Tenebrio* pupae.[a]

Neotenin: Oil	Average Scores
1:0	6.5, 7.8, 9.5
1:4	4.5, 4.8
1:16	2.0, 2.5
1:64	0.8, 0.8

[a] Data after Wigglesworth (1958).

Additional experiments can be conducted in which various insects are tested by techniques similar to those described by Wigglesworth (1958) on terminal instar *Rhodnius* nymphs. Neotenin extracts can be injected to produce gigantic extra instars in some cases. Terminal instar larvae of *Tenebrio molitor* or of various trichopteran and lepidopteran larvae are convenient subjects.

Other assay techniques are available which give more refined results. For example, Karlson and Nachtigall (1961) have defined a *Tenebrio* unit (TE) in terms of juvenile hormone activity. Gilbert and Schneiderman (1960) have described a polyphemus bioassay injection method for *A. polyphemus* which employs eight easily recognized structural features. After a given injection and molt, each structure is rated from zero (all adult) to five (all pupa) and the points are totalled; the assay animal can receive a score of zero to forty. The scores are adjusted for inactivation of the hormone in the pupal system and for activation of the prothoracic glands; corrections are based on time of molting following removal of the pupa from 6° to 25° C.

The problem of lack of comparability of the results of various assay methods has been discussed by Gilbert and Schneiderman (1960) and Wigglesworth (1963).

3. COMPOUNDS HAVING ACTIVITY

A number of compounds with various degrees of juvenile hormone activity have been reported. Schneiderman and Gilbert (1964) used an assay method in which cecropia oil was assigned a value of 1000 and found that the relative activity assay of commercial farnesol was 16, of all-*trans* farnesol 140, of farnesyl-*o*-methyl ether and of farnesyl diethylamine about 1400. For the convenience of those desiring to experiment with these compounds, sources of supply are listed in the Appendix. There is evidence that dilutions of these compounds in distilled water or 10% ethanol rather than in oil may be more suitable. Before using the above compounds, the 1964 review by Schneiderman and Gilbert and various references cited therein should be consulted.

If neotenin analogues can be obtained it is instructive to test them using the *Tenebrio* bioassay given above as well as on a series of other insects previously tested with neotenin extracts. In particular, the effects of the terpene derivatives mentioned above, which show greater activity than extract of cecropia, may be effective in producing extra-instar giants.

CHAPTER FIVE

Behavior

The study of insect behavior has long been an important, active area and one holding considerable fascination. No biologists should forego the pleasure of reading the classic observations of Jean Henri Fabre (1876–1904), the backyard entomologist *par excellence.* With the development and utilization of new techniques, Fabre's original natural history approach to behavior has evolved into an experimental science. The literature on behavior has become very extensive; reviews and summaries have been presented by Schnierla (1953), Carthy (1958) and Fraenkel and Gunn (1940) and should be consulted for an overview of modern work, which seeks to relate behavior to physiology, genetics, ontogeny, phylogeny and ecology. The exercises offered in this chapter are but a sampling, intended only to serve as an introduction to an experimental approach to the study of behavior.

Studies with caddis fly larvae provide information on fixed and/or modifiable behavior concerned with the selection and manipulation of materials in the environment. The examples drawn from studies of mating behavior of *Drosophila* and certain singing Orthoptera illustrate the primary importance of such behavior in reproductive isolation; these two examples also demonstrate the roles of communication by sight and sound between members of the same and of the opposite sex, as well as between species, in mating behavior. The section on web building and orientation in spiders provides comparative data on a group of related arthropods.

A. CADDIS FLY CASE-BUILDING BEHAVIOR

Larvae of the insect order Trichoptera provide particularly interesting examples of both fixed and modifiable behavior. Of the nineteen families of North American Trichoptera, the larvae of all but the Rhyacophilidae construct cases or nets. Larvae are found in almost all unpolluted freshwater habitats and a great many species have overwintering generations.

Most species are easy to collect and rear in the laboratory and are ideal for behavioral studies. Their particular advantage is the fact that the results of experimental manipulation—the case or net—can be preserved as a permanent record.

Most of the research on caddis fly case-building behavior has been either descriptive and primarily concerned with normal patterns (Vorhies, 1905; Murphy, 1919; Lloyd, 1921; Wessenberg-Lund, 1943; Ross, 1944; Nielsen, 1948; Flint, 1956; Hanna, 1959) or directed toward establishing the range of materials that larvae are capable of manipulating ("behavioral plasticity") (Marshall and Vorhies, 1905; Gorter, 1928; Frankhauser and Reik, 1935; Webster and Webster, 1943; Dudziak, 1951). Some general patterns have emerged from these studies (Hanna, 1960) and have been related to phylogeny (Ross, 1956) and ecology (Milne, 1943; Cummins, 1961) of the group. Behavioral characteristics have also been utilized in the systematics of Trichoptera (Wiggins, 1960).

A great deal of information on normal and modified case-building patterns in the Trichoptera remains to be gathered before the ontogeny, phylogeny and ecology of the group can be definitively evaluated on the basis of behavior.

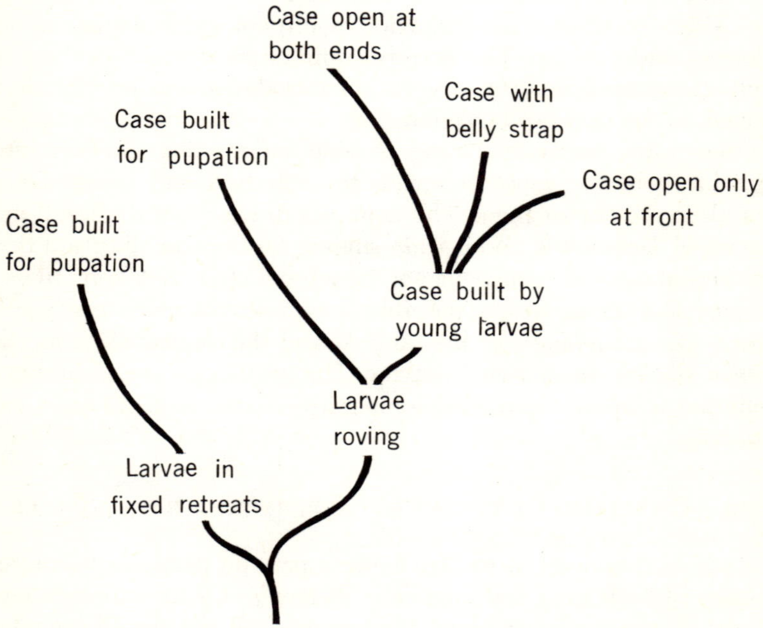

Figure 5.1. Evolution of case building in Trichoptera. (Based on Ross, 1956)

In his excellent work on trichopteran phylogeny Ross (1956) discussed the evolution of case type within the order; Figure 5.1 shows the probable phylogenetic relationships of the caddis flies. One line led to larvae that construct fixed retreats (see Chapter Two, Section C), the other to free-living larvae that build tortoise-shell-like pupal cases and eventually to larvae that bear portable cases. As Ross points out, a single mutation could have resulted in shifting the case-building act from mature larvae to first instar individuals.

1. CASE-BUILDING PATTERNS

As a preliminary to studies on case building in Trichoptera, it is useful to observe basic patterns of case building and to test the degree to which these patterns can be modified.

MATERIALS. Cultures of trichopteran larvae, preferably species in the family Phryganeidae; culture vessels with means of aeration.

Miscellaneous food materials collected from the natural habitat of the caddis fly larvae under study, such as *Oedogonium, Cladophora,* partially decomposed *Salix, Cornus* or *Alnus,* wood fragments or other vegetal material. Unless the food requirements of the species under study are specifically known (see Chapter Six, Section B 3) a range of materials must be supplied. Stream forms may require a continuing supply of field-collected diatoms, such as *Gomphonema, Meridion* or *Amphora.*

Miscellaneous natural case materials such as leaf and stem fragments for studies with phryganeid species, sand and silt particles for mineral case builders. Care should be taken to separate mineral particles used into size fractions (see Cummins, 1961, 1962, 1964). Miscellaneous "foreign" materials for case-construction experiments, such as wax paper, polyethylene sheets and tubing, aluminum foil, and the like.

Vials; 70% ethyl alcohol; labels.

METHODS. Observations of case building should be made at timed intervals and recorded in tabular form. The temperature and light conditions under which observations were made must be noted. Critical stages of the larvae and cases should be preserved in 70% ethyl alcohol and appropriately labeled. When feasible, a tape recorder and photographic equipment are useful for permanently recording data.

Phryganeid larvae, such as *Ptilostomis ocellifera* (Walker) or *Phryganea sayi* Milne are suitable for observing normal patterns of case construction in terms of selection and manipulation of materials and elaboration of the typical case form. Eject a larva from its case by pressing a blunt probe through the posterior case opening; replace larva and case in the rearing vessel. Note that the larva finds, recognizes and returns to its case. Eject another larva in the same manner, but this time offer it a selection of empty cases, including its own. Can the larva now find and recognize its own case among those presented? Again eject a larva and place it in a vessel containing leaf and stem material. It will usually construct within a few hours an irregular, rough case similar in length to the original one. During more orderly building, soon initiated at the anterior end, case length is approximately maintained. The larva cuts off posterior sections of rough case after each period of new construction at the anterior end until a case similar in form to the original has been completed. The entire process requires one to three days.

When the anterior half of a larval case is removed, either irregular or finished-type regular construction may result. The smaller the removed portion, the greater the tendency toward regular construction. If repair of the removed portion is irregular, the regular form is achieved by new construction at the anterior end as described above.

Figure 5.2. Natural and reconstructed trichopteran cases compared. (A) natural and (B) reconstructed cases of *Ptilostomis ocellifera.* (C) natural and (D) reconstructed cases of *Phryganea sayi.*

Figure 5.2 shows natural and reconstructed cases of *Ptilostomis ocellifera* and *Phryganea sayi.*

Cummins (1962, 1964) investigated the case-building behavior of two species of limnephilids–*Pycnopsyche lepida* and *P. guttifer.* Studies of case construction initiated by shortening cases of *P. lepida* and placing them on substrates composed of mixtures of various sizes of mineral particles, revealed a strong selection for particles approximately 1 mm. in diameter. Table 5.1 shows the relative abundance of different mineral particle sizes and organic fragments in natural *P. lepida* cases; the dominance of the 1 mm. particle sand grains in experimentally induced case construction is shown in Table 5.2. When larvae were observed during reconstructive case building the selection of 1 mm. particles was further clarified. As given in Table 5.3 many more particles were handled by

Table 5.1 Mineral and organic components of five full grown *P. lepida* larval cases [a]

Case Length, mm.			Mineral Portion; No. of Particles of Given Size, mm.					Organic Portion; No. of Pieces (Greatest Length, mm.)					
Total Length	Sand Length	Organic Length	¼ and Finer	½	1	2 and Larger	Mineral Total	½ and Less	1–2	2–4	4 and Longer	Organic Total	Grand Total
22	16	6	39	13	37	19	108	0	36	29	16	81	189
22	17	5	111	48	76	46	281	1	25	27	5	58	339
20	18	2	25	42	109	24	200	Not measured					200
20	20	0	36	95	193	33	357	Not measured					357
20 [b]	17	3	83	87	212	18	400	0	9	8	4	21	421
Anterior half			19	50	134	9	212	0	0	0	0	0	
Posterior half			64	37	78	9	188	0	9	8	4	21	
Average numbers as %			18	18	39	9	84	1	7	6	2	16	100

[a] Cummins, 1964.
[b] Anterior and posterior halves of case analyzed separately.

Table 5.2 Particles utilized by full grown *P. lepida* larvae, in shortened cases, on a mixed substrate of ⅛–4 mm. particles (30 ml. each)

Duration of Experiment, days	Length of Case Added, mm.	Number of Particles Utilized					
		⅛ mm.	¼ mm.	½ mm.	1 mm.	2 mm.	4 mm.
1	10	7	3	25	48	7	0
1	10	1	1	3	30	25	0
4	7	5	19	35	44	6	0
5	4	4	10	10	19	10	0
3	2	0	2	6	6	2	0
1	2	1	1	3	9	4	0
1	7	5	7	25	54	12	0
6	6	5	7	18	25	11	0
Totals		28	50	125	235	77	0
% total		5.4	9.7	24.3	45.6	15.0	0

Table 5.3 *P. lepida* case-building behavior; observations on full grown larvae, in shortened cases, placed on a mixed substrate of ⅛–4 mm. particles (30 ml. each)

Number of Particles Handled by Larva	Number of Particles Added to Front of Case	Number of Times Particles Held in Various Positions against Leading Edge of Case Prior to Addition	Length of Observation, hr.
5	5	not recorded	½
29	4	not recorded	1
42	4	11	2¼
10	3	16	1½
Totals 86	16	27	5¼
Averages 17	5	14	

larvae than were incorporated into the case; the 1 mm. particles were most frequently incorporated. Each particle added to the anterior margin of the case was "fitted" into position as indicated by the considerable amount of manipulation that preceded actual incorporation of a particle into the case. It seems that the smooth, regular, masonry-like appearance of *P. lepida* cases (Fig. 2.22) is due to selection of sand grains of rather uniform size and a careful fitting process.

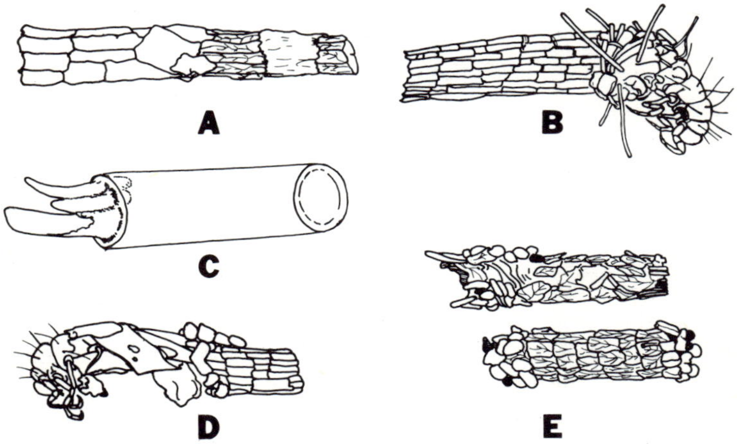

Figure 5.3. Results of experimental manipulation of case building by phryganeid larvae. (A) *Ptilostomis ocellifera,* wax paper added to posterior section of original case; (B) *Phryganea sayi,* sand grains and wire scaffolding used for repair; (C) *P. ocellifera,* leaf fragments added to a section of plastic tube; (D) *P. sayi,* posterior section of original case repaired with aluminum foil; (E) *P. sayi,* sand grains used to repair shortened sections of original cases.

Based on observations, such as those outlined above, of the fixed behavior patterns of normal case construction, further experimental manipulation can be employed to elucidate the modifiability of the patterns.

If phryganeid larvae are ejected from their cases, or if portions of their cases are removed, and the larvae placed in vessels containing a variety of mineral and organic materials from the natural habitat, the substances normally utilized in construction are invariably selected. If, however, such larvae are placed on substrates consisting entirely of foreign materials never encountered in natural situations, construction results nevertheless (Fig. 5.3). The similarity between the case reconstructed with foreign materials and the natural case depends upon the physical similarity between the foreign substances and material normally utilized. As several workers have shown (Frankhauser and Reik, 1935; Dudziak, 1951), case building may proceed even with very unusual materials such as aluminum foil (Fig. 5.3D). Construction with sand grains by phryganeid larvae is enhanced if some sort of scaffolding material is provided. Dudziak (1951) supplied pine needles; pieces of fine wire were used in the construction of the case shown in Fig. 5.3B.

Studies on the materials selected for case construction can be facilitated by inducing larvae ejected from their cases to establish residence in polyethylene or glass tubing of the appropriate inside diameter. In this way the exact amount of material added in a given time can be removed (Fig. 5.3C).

2. ONTOGENY OF CASE BUILDING

Although zoologists have focused considerable attention on morphological developmental changes of invertebrates, little work has been done to delineate developmental behavior patterns (for sequential morphological patterns see Fox and Fox, 1964: 210–283 and Johannsen and Butt, 1941). Many behavioral sequences are as stereotyped and predictable as the better known morphological stages. The normal pattern of trichopteran case building often follows a definite developmental sequence involving a change both in case form and in the selection of building materials. Since many such behavioral changes occur during a given instar, it is clear that they are not just secondary effects of an altered morphology.

Ontogenetic changes in case form of species of Limnephilidae and Molannidae have been discussed by Wessenberg-Lund (1943) and a very detailed study of the microcaddis flies (Hydroptilidae) was completed by Nielsen (1948). Hanna (1957, 1959) observed that the cases of five species of trichopterans in three families increased in size paralleling the growth of the larvae and that the length and diameter of vegetal fragments utilized by two species of limnephilids also increased with larval age. Tindall (1960) presented evidence that the diameter of stem material employed is measured by the larvae by use of the labral notch (Fig. 1.5E, n). Since the labrum, along with other sclerotized parts, increases regularly in size with each molt, this habit would account for the increase, as the larva grows, in diameter of stem fragments used.

Cummins (1964) showed that the limnephilid *Pycnopsyche lepida* (Hagen) changed its pattern of case construction and selection of materials during the terminal instar. The concomitant shift in microhabitat associated with the change in case building results in a lack of contact between *P. lepida* and the closely related *P. guttifer* (Walker) occurring in the same stream.

Materials. Since limnephilid caddis fly larvae abound in many lentic and lotic habitats, the exercise below is designed for work with them; the fact that the eggs are terrestrial facilitates collection and handling. Eggs of species belonging to other families, although deposited beneath the water surface, are also suitable (for descriptions see Ross, 1944). Limnephilid eggs can readily be found in the soil or on vegetation close to the bank and enclosed in small gelatinous masses. These masses should be kept in a moist atmosphere until ready for use, at which time they can be submersed in aerated culture vessels. Additional materials are as outlined for the analysis of case-building patterns above.

Methods. Although the description that follows applies specifically to the genus *Pycnopsyche,* it is applicable with slight modifications to most limnephilids.

When the eggs are placed in water they begin hatching in from two to six days. As the larvae hatch they should be removed in groups of ten to additional culture vessels. Observations should be made and detailed notes kept concerning hatching, molting, case building and feeding. After the larvae have left the gelatinous matrix of the egg mass, some individuals can be introduced into culture vesssels containing equal amounts of two or more building materials, thus making possible observations on the importance of selection in case building. Information about changes in case building during larval growth (both inter- and intra-instar changes) are of critical importance.

Discussion. The act of case building seems to be under fairly direct genetic control, as exemplified by the fact that none of the environmental influences studied by Cummins (1962, 1964) stopped case building completely. On the other hand, it is quite apparent from the studies considered above that the manipulation of case-building materials is learned by larvae and that the materials utilized to a greater degree, and case form to a lesser degree, are subject to environmental pressure.

Tindall (1960) described an interesting situation encountered in the larvae of the leptocerid *Triaenodes bicolor* Curtis, of which there are apparently two genetic strains—one of which constructs sinistrally, the other dextrally wound cases. Larvae found in one case type continue to construct according to the same spiral pattern regardless of experimental manipulations.

The actual neuromuscular basis of case building in caddis flies is just beginning to receive attention. Merrill (1964) found that control of case length in certain phryganeids and limnephilids is mediated through receptors associated with the anal hooks. When the hooks and associated sensory setae are removed, the larvae produce abnormally long cases (Frontispiece).

Experimental studies of trichopteran case building have thus far been confined to relatively few species—primarily in the families Phryganeidae, Limnephilidae and Molannidae (see review by Hanna, 1960). A host of intriguing problems remain to be investigated, among them the effect of stream current or lake wave action on the selection of particles to be added laterally to flanged cases (*Neophylax, Molanna, Goera*), and the selection and sizing of fragments used by *Brachycentrus* larvae to construct its tapering case, square in cross section. The behavior of the common snail-case builders (*Helicopsyche*) has not been studied despite the fact that its case is perhaps the most unusual in the order and poses some interesting evolutionary questions.

B. COURTSHIP BEHAVIOR OF *DROSOPHILA*

The most complete summaries of courtship in *Drosophila* are those of Spieth (1952), which describe the mating behavior of the majority of species, especially those in the Western Hemisphere, and of Brown (1964), an exhaustive survey of the mating patterns of the species of the *D. obscura* group. The basic account of courtship of *D. melanogaster*, generally the most easily obtained species, given below is based on Bastock and Manning (1955).

The male approaches the female and taps her on the body with his foreleg. If the female moves away the male may or may not pursue her. If she remains in place, he stands facing her and extends the wing nearest the female, vibrates it for a few seconds and then closes it again. This phase is often repeated several times while the male moves around the stationary female. If the female changes her position during vibration, the male may vibrate the other wing if it becomes the one nearer her. Eventually the male will position himself behind the female, extend his proboscis and lick the female genitalia. Finally he will attempt to mount the female, a procedure in which she must cooperate by spreading her wings, otherwise the male will be unable to maintain position and will fall away. These steps may be repeated several times but the sequence always is substantially the same. The male may have to begin again at any point in courtship if the female is unreceptive, or the entire procedure may be terminated at any stage for one reason or another. Mating usually occurs within a few minutes but the time is variable.

The elements of courtship differ markedly from species to species (Spieth, 1952) and the sequence of events appears to be characteristic of the species, serving as an isolating mechanism among closely related sympatric species.

MATERIALS. Cultures of *Drosophila melanogaster* and of at least one other species; it is essential to have aged males and virgin females; flies at least five days old are best. Clear plastic boxes no more than one-quarter of an inch deep and just large enough to fill the field of the dissecting microscope at lowest power serve best as observation chambers; shell vials stoppered with cotton may also be used. Dissecting microscopes; culture bottles; *Drosophila* food.

METHOD. For each set of flies to be observed prepare a record sheet to show the sequence and time of each event of courtship. Behavioral phases to be included are (1) **orientation,** when the male circles the female, (2) **tapping,** when the male taps the female with his foreleg, (3) **vibrating** of the male's wing, (4) **licking,** when the male licks the female genitalia, and (5) **mounting,** when the male attempts copulation. Observation and recording is best done by two persons working as a team, one using the microscope, the other timing and recording.

Prepare the observation chamber by placing a small bit of *Drosophila* food in the bottom of it. Introduce three pairs of unanesthetized flies and place the chamber under the microscope for observation. Particular attention should be paid to the timing and the sequence of events for each species observed. From the data obtained prepare a graphic representation of mating behavior (Bastock and Manning, 1955).

C. COMMUNICATION BY STRIDULATING ORTHOPTERA

While expensive, highly specialized equipment for recording and analyzing insect songs is required for refined research (see Alexander, 1957b and Haskell, 1961: Chapter 1), much can be accomplished with inexpensive tape recorders. In recent years it has been demonstrated that low-priced recording equipment can be used to obtain tapes of animal sounds having sufficient fidelity to elicit species-specific responses on playback (Legler, 1964). Many excellent discussions of insect songs and the mechanisms of sound production by insects are available (Comstock, 1948; Pierce, 1949; Dethier, 1953; Haskell, 1961) and a valuable bibliography has been compiled by Frings and Frings (1960).

Crickets (Grylloidea) produce sound by stridulation, using structures on the forewings (tegmina). The **file,** a modified vein (Fig. 5.4) on the anal lobe, of one tegmen is drawn across the rigid marginal edge (**scraper**) of the other forewing, causing it to vibrate; the sound is augmented particularly by a special resonating area termed the **mirror** by Haskell (1961) or, more appropriately, the **tympana** by Comstock (1940).

The object of investigation in the present exercise is the role played by stridulation in the reproductive behavior of field crickets.

MATERIALS. See Chapter Two; the recordings of stridulation made for the systematics exercise with *Gryllus* will be used.

METHODS. Silence (destridulate) males, place them in cages with a female of the same or a different species and play back tapes of male songs of various species. Destridulation can be accomplished by clipping off the forewings of male crickets or by coating the files and scrapers with paraffin brushed on with an insect pin dipped into melted wax.

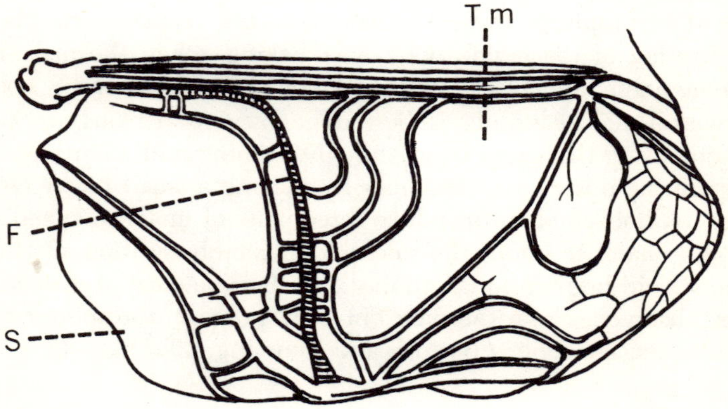

Figure 5.4. Posterior (anal) lobe of tegmen of *Gryllus* species. F, the file, a thickened vein with roughened surface; S, the scraper, the rigid membrane at the anal margin; Tm, tympanum, a membranous resonating area.

To establish a behavioral base line, a male cricket can be placed in a cage containing a female of the same species, along with females of one or more other species. To facilitate identification, the females of the various species can be marked with dots of nail polish on the tegmina. It will be observed that the song of the male attracts only the female of its own species. Conversely, a female can be encaged with males of her own and other species; it will be seen that she is attracted only by the stridulation of her own kind.

That stridulation characteristic of the species is essential to recognition by crickets of its own species and to initiation of courtship behavior is readily demonstrated by repeating the base line experiments but using destridulated males. In the absence of the species-characteristic song, the females are at a loss to recognize their males. Even when a female is caged only with a destridulated male of her own species, with no other males present, her interest in him is not aroused. Need one wonder at the successes of the troubadors of old?

More elaborate experimentation is possible and Haskell (1961) should be consulted if it is desired to pursue this topic.

D. WEB BUILDING AND ORIENTATION IN SPIDERS

In her charming little book, *The Web of the Spider,* Lougee (1963) discusses aspects of web-building behavior. As is true of case type in Trichoptera, web type in spiders is diagnostic of family, genus or species. A number of references dealing with spiders are available and should be consulted (Comstock, 1948; Gertsch, 1949; Millot, 1949; Kaston and Kaston, 1953; Emerton, 1961; Fox and Fox, 1964). The key in Kaston and Kaston (1953) is particularly useful in connection with web building.

Materials. Lougee (1963) describes a method for collecting suspended webs, especially the webs of the orb-weaving spiders (Araneidae), using sheets of heavy black construction paper (8½ inch × 11 inch or larger), black lacquer paint or liquid black latex, clear and white lacquer sprays, a fine brush and wire hoops fashioned from heavy wire (such as coat hangers) of various suitable diameters. Needed also are ring stands and clamps, masking tape and a light source such as a microscope lamp.

Methods. The spider is removed from the web and preserved in 70% ethyl alcohol to which a little glycerin has been added. The web is sprayed *in situ* with white enamel applied in several very light coats with the spray can held a foot or more from the web. The enameled web is picked with a piece of heavy black construction paper by lifting the paper into the web from beneath and behind. Cut the structural strands around the margins of the paper and apply several light coats of clear lacquer spray. Such preserved webs properly associated with the spiders constructing them are useful for laboratory study and for reference collections. In some cases structural detail of the web can be seen better if clear spray is used instead of the white enamel, but only experience can guide the decision.

Examine the webs collected and identify the spider species, using the key given by Kaston and Kaston (1953). The structure and building of a typical orb web (Araneidae) is shown in Figure 5.5. Orb web spiders use two kinds of silk—dry and viscous threads. Note the central hub composed of irregular cells and constructed with dry threads. An attachment zone, also made of dry silk, is seen around the hub. The main body of

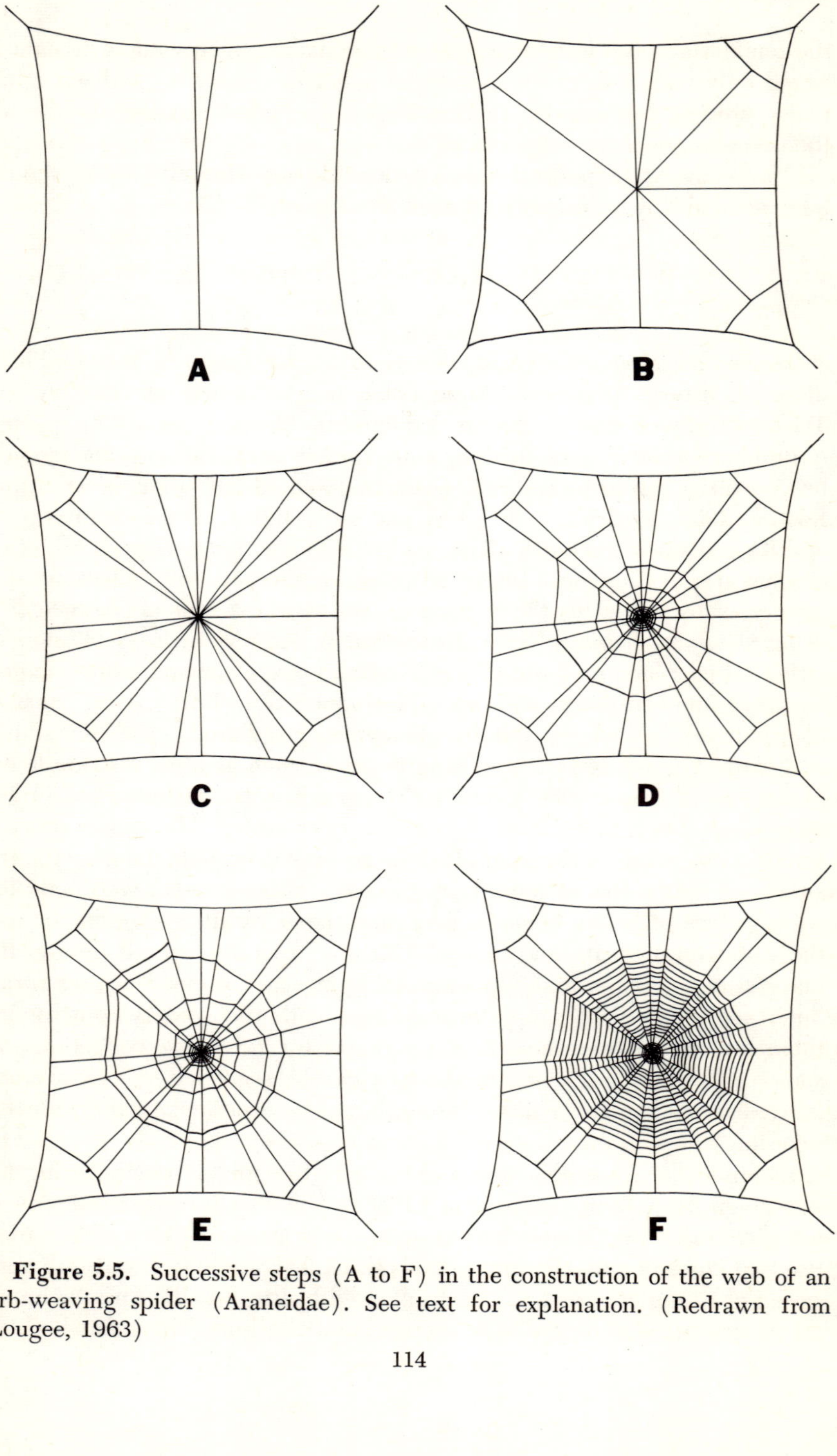

Figure 5.5. Successive steps (A to F) in the construction of the web of an orb-weaving spider (Araneidae). See text for explanation. (Redrawn from Lougee, 1963)

the web is the spiral zone, used to trap prey and made up of viscous threads.

Web building begins with the foundation laid down around the margins and anchored. Primary radial threads, spun double as the spider retraces her path, are next attached across the area and connecting the foundation lines. The number of such double-spun primary radii tends to be fairly constant for a species. Secondary single strand radii are next spun across the network and all radii are fastened at the hub. Finally the viscous spiral is spun.

For behavioral studies, the collecting procedures outlined above are not suitable. The live spider and her unsprayed web must be transported to the laboratory. Transfer the spider to a small container such as a corked vial, plastic box or capped jar. Pick up the web on a wire hoop of suitable diameter; if bits of vegetation must be included to maintain the integrity of the web they can be cut free and affixed to the hoop with masking tape.

In the laboratory, support the hoop by a clamp on a ring stand and adjust so that the original plane of orientation is restored. Release the spider on her web. Notice the place on the web used by the spider at rest. Allow a small insect such as a cultured *Drosophila* to become entangled in the web and observe the spider's behavior associated with preparation of her prey and with feeding.

Damage the web slightly and observe repair behavior. After a spider has been in residence for some time it may be possible to observe the construction of a new web following the nearly total destruction of the existing one.

Change the plane of orientation of the web by adjusting the ring stand clamp and notice the changes in behavior resulting. Damage the web slightly and observe whether the new plane of orientation alters the previously observed repair behavior.

Place one of the hoop-supported webs in the dark or under conditions of subdued light. Illuminate it strongly from one side with a microscope lamp. Elicit prey preparation, feeding and repair behaviors and observe. Move the light source to the opposite side and again elicit behavioral responses; note any change in behavior. Repeat with the light source above the web and then with the light source below the web. It will be noted that the movements of the spider are oriented by the position of the light source, which appears to be of greater importance than structural landmarks on the web. Verification can be obtained by keeping the light source constant and rotating the hoop 90°, 180° and 270° on the ring stand clamp.

If species of the family Agelenidae (see references above) can be obtained, the orientation experiments performed by Bartels and Baltzer (1928; described by Carthy, 1958) can be duplicated. The agelenid spiders construct a sheet web with a funnel shaped tubular retreat. In some species the retreat is at one corner of the net. Bartels and Baltzer found that if the web is rotated with respect to a unidirectional light source, or if the light is moved, the spider is unable to find the retreat. They attributed this loss of orientation on the web to the importance of light response.

Similar studies on light orientation in spiders can be achieved by painting over the eyes with black lacquer or liquid latex applied with a fine brush. The behavior of partially or completely blinded individuals may be compared to normal behavioral patterns.

CHAPTER SIX

Ecology

As Odum (1959, 1963) has pointed out, natural history gave birth to ecology when investigators asked not only what kinds of animals inhabit a given environment, but also how many. Unfortunately, this respectability gained by natural history through quantification has often been paid for by a loss of detailed observation—particularly in the field.

Ecology, as the study of the relationships of organisms to the environment—physical, chemical and biological—has achieved considerable prominence in recent years, due primarily to the fact that many practical problems of human health and welfare require basic ecological data. The broad scope of ecology can be appreciated only with the realization that individual organisms, populations, communities, ecosystems and the biosphere as a whole all are under scrutiny by ecologists. It is no wonder that subjects such as physiological ecology (Dill, Adolph and Wilber, 1964), behavioral ecology (Klopfer, 1962) and ecological genetics are becoming commonplace in textbook catalogues and among course listings.

Animal ecologists have found insects to be ideal subjects both for field and laboratory investigations in both the areas of research that currently dominate animal ecology—population dynamics and energetics (see Slobodkin, 1961; Cragg, 1962; Hazen, 1964).

A. POPULATION ANALYSIS

Populations long have been of interest to the ecologists and more recently have attracted the attention of the geneticist and the systematist. It is useful, therefore, to study wild populations and to define as many of their parameters as possible; it is dangerous to extrapolate conclusions from laboratory situations to wild populations. Many successful capture-recapture experiments have been recorded using vertebrates, but such studies with insects are rare. The finest work in this regard has been that of Dowdeswell *et al.* (1940) using the satyrid butterfly *Maniola jurtina*

(Linné) and those of Ehrlich (1961a, 1961b) with the nymphalid *Euphydryas editha* (Boisduval).

1. FIELD POPULATION CENSUS

This and following experiments demonstrate some of the parameters of natural populations: the effective size of a population at a given time, the total size of the population, the life span of the individual, sex ratios and ratios of polymorphic forms. Such parameters can be calculated directly if the sampling of the population has been adequate. For example, it is possible to estimate the turnover rate of individuals from the data collected. Dispersal of single individuals and/or of the population as a whole may be demonstrated by plotting recaptures on maps.

The method for the bulk of the experiments is substantially that of Ehrlich and Davidson (1961), and the study by Lawson *et al.* (1963) suggested the exercise using the ultraviolet light trap.

MATERIALS. Compass; marking stakes to delimit individual areas; fifty (or 100) yard tape measure; heavy twine to mark borders between individual areas; paper for maps; right triangles; scaled ruler; drawing pencils. Each student should be supplied with the following: aerial insect net; at least one pair of spatula forceps (stamp tongs); prenumbered glassine envelopes; felt marking pencil, such as the "Magic Marker"; glass plate; record paper; pencils.

METHOD. *Selection of the study area.* It is vital to the success of this experiment that a small, localized population of some insect be chosen. A large area cannot be adequately sampled by a class and, if the insect selected is not a localized one, immigration and emigration will result in a swamping of the population under study and loss of marked individuals. For reasons enumerated below a colony of the orange sulfur butterfly *Colias eurytheme* (Boisduval) in an isolated alfalfa field is ideal for this type of study. Dimorphism in this species enables easy identification of the sexes (see Fig. 6.1A, B) and in most areas the females are polymorphic—some being orange, others white.

Mapping the study area. Once the study area has been selected it should be divided, using compass and tape measure, into sections fifty to a hundred yards on a side. The boundaries should be delineated by marker stakes with twine strung between them. These sections should then be plotted on a master map with an appropriate scale (1 inch to 10 yards would be convenient). Notable features of each quadrant, such as trees or gullies, should be noted on the map and each quadrant assigned a code number. The sections are then assigned to individual students who are responsible for them for the duration of the study. Each student should make copy maps of his own area from the master map on which to record his daily captures.

Records. Accurate records are essential to the success of this experiment, and their importance cannot be stressed too heavily. Each student should prepare, or have provided, record sheets giving the following information as a minimum: student's name, section number, species of insect sampled, specimen number, sex of the specimen, date of original capture, date(s) of subsequent recapture(s), section number(s) in which recapture(s) are made.

It should be noted that specimens originally marked by a student remain his responsibility throughout the experiment; cooperation is accordingly necessary among students in reporting recaptured specimens to their original "owners."

Collecting. Students should collect specimens *only* within their assigned areas, netting each individual in the area and placing each in a separate prenumbered glassine envelope. Great care must be exercised in removing the butterfly from the net to the envelope to guard against injuring the specimen. The butterfly should be held lightly but firmly by the thorax through the net, the net opened carefully, and the butterfly grasped by all four wings with spatula tipped forceps. Now the butterfly may be placed in the envelope and the envelope closed but not sealed. The envelope should then be put in a cool place until time for removal and marking. The place of capture of the specimen should be noted, according to the number on the envelope, on the student's section map for later transference to the master map.

Marking. At the conclusion of the collecting period all captures should be brought to a central meeting place for marking and recording. After the first day specimens must be checked carefully for previous markings and reported to the original captor.

The suggested marking system is based on the 1–2–4–7 system used to record data on punched cards. As shown in Figure 6.1C, the *1* position is at the apex of the forewing, *2* at the anal angle of the forewing, *4* at the apex of the hindwing, and the *7* position is at the anal angle of the hindwing. All marks should be applied to the under surface of the wings. The left side of the insect (the right when viewed from the under side) is the *units* side, and the right side is the *tens* column. If the size of the population is large, the discal cells of each wing may be used for a *hundreds* designation by assigning the right forewing cell the *1* position, the right hindwing cell the *2,* the left forewing cell the *4,* and the left hindwing cell the *7* designation. By this method it is possible to mark at least a thousand specimens with a single color of ink.

The combinations of numbers for the marking procedure are as follows: for 1 mark *1* only, for 2 mark *2* only, for 3 mark *1* and *2*, for 4 mark *4* only, for 5 mark *1* and *4*, for 6 mark *2* and *4*, for 7 mark *7* only, for 8 mark *1* and *7*, for 9 mark *2* and *7*, and for 10 mark *1* in tens column, no marks in units column. The key for marking and sample marked individuals are shown in Figure 6.1.

The marking of a specimen is ideally a two person task. One grasps the specimen gently but firmly by the thorax with his left hand (if he is

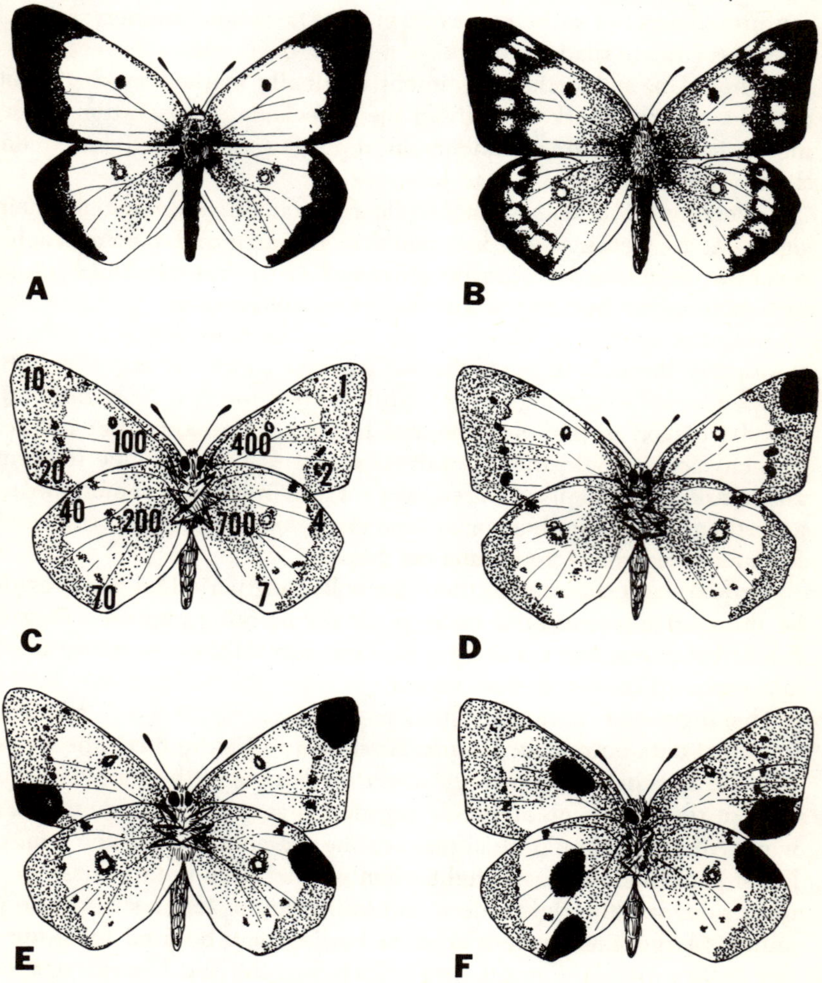

Figure 6.1. *Colias eurytheme* (Boisduval). (A) upperside of male; (B) upperside of female; (C) key to marking code on underside; (D) a male marked as number 1; (E) a male marked as number 25; (F) a male marked as number 376.

right-handed), and the other slips a glass plate between the wings and manipulates the wings with forceps while applying the markings. The first person then makes the appropriate marks on each wing with the felt marking pencil. When the markings have dried, the butterfly is returned to the glassine envelope. After the marking has been completed all specimens should be released in the center of the area in which they were captured.

As has been noted before, recaptured specimens are the key to the success of this experiment. When a marked specimen is recaptured it should be returned to the person who originally marked it, who will then record its recapture date and the area in which it was taken. Then the specimen is returned to the person who last captured it, and it should be released in the area of the last recapture.

HANDLING THE DATA. *Size of population.* The classic Lincoln index (Lincoln, 1930) for estimating population size may be used in this experiment. It is based on the ratio of marked to unmarked individuals in an experimental population and may be expressed by

$$P = \frac{(I_m \quad C_t)}{I_r}$$

where P is the size of the population; I_m is the number of individuals marked; I_r is the number of individuals recovered, and C_t is the total number of individuals captured, both marked and unmarked. Three assumptions are inherent in the Lincoln index: (1) marked individuals disperse at random throughout the population, (2) there is no immigration into or emigration from the population, and (3) there is no tendency for marked individuals to seek or unduly avoid recapture. This index should be calculated for each collecting period; the results of later calculations will show the true population, since the estimate tends to level off at one value.

Sex ratio and ratio of polymorphic forms may be calculated directly from the capture totals for each day, as well as for the entire study. Comparison of the ratios obtained for each collecting day will indicate whether or not the flight time of the sexes and/or polymorphic forms differs.

Life span may be estimated by noting the earliest and latest dates of capture for marked specimens, assuming no selective disadvantage to marked individuals. Since specimens have generally been flying for some time before original capture and usually fly after last capture and before death, the longest life span obtained is probably the best estimate.

Turnover of individuals is related to life span as well as to immigration and emigration. It may be calculated by the ratio of new captures in a day's collecting to total captures.

Dispersal is a parameter of the population which may be shown by noting the places of capture and recapture of individuals. These data should be shown on the master map and the dispersal of individuals should be compared for information as to whether the dispersal of the population is of a random nature or whether there are intrinsic factors involved in the mobility of the population.

Analyses of data obtained in a similar way are given by Ehrlich (1961a, 1961b).

2. POPULATION SAMPLING BY CHEMOATTRACTION OF SATURNIID MOTHS

The following experiment serves several purposes: it demonstrates the importance of the sexual attraction of the female saturniid moth; it will afford an estimate of population size for the moth studied, and specimens can be collected for extracting juvenile hormone for use in experiments outlined in Chapter Four. For the last purpose the species of choice is the cecropia moth *Hyalophora* (*Hyalophora*) *cecropia* (Linné), but any saturniid moth will suffice for the population study.

It is necessary to collect cocoons so that only virgin females are used for this experiment. Females taken at lights have almost always mated and will not be of further interest to males. When the moths begin to emerge under natural conditions, many will undergo eclosion at the same time, and it is essential to separate males from females as quickly as possible. The males should be marked according to the 1–2–4–7 system detailed above (Fig. 6.1) and stored until needed for the experiment.

A warm, relatively windless evening should be selected. The marked males should be released at points along a study area gridded at one-quarter or one-half mile intervals, marked on a U.S. Geological Survey Quadrangle map of the area in question, preferably one of the newer 1:24,000 (7.5 minute) maps. The points of release and the numbers of individuals released should be plotted on the map. A cage with a single female is placed near the center of the gridded area and the cage checked at intervals for attracted males. With the exception of one species it will be necessary to check for males during the night time hours, so the cage should be placed in an accessible location. The approximate times of attraction are given in Table 6.1 for the more common saturniid moths of the eastern United States.

All males attracted to the cage should be collected and checked for number markings. Knowing the points of release of the recaptured marked males will give the distance over which males were attracted. Plot the farthest points from the female in all directions from which males were attracted. Direction and velocity of wind during the experiment should be marked on the map. By taking the number of males released within the area of attraction, an expected number of males is obtained. Those expected males that did not appear at the cage with the experimental female were probably visiting other females that had emerged within the study area. From the number of unmarked males collected it is possible to derive an estimate of the size of the wild population, again assuming no selective advantage or disadvantage to marked males, by using the formula

$$\frac{M_e}{M_o} = \frac{P_e}{P_o}$$

where M_e and M_o are the marked individuals expected and observed, respectively, and P_e and P_o are the wild (unmarked) individuals expected and observed. The equation is solved for P_e, giving the expected numbers of males in the wild population, and since males and females emerge in about equal numbers, doubling the obtained estimate gives an estimate of the total wild population. Density of moths per square mile can then be easily calculated.

Table 6.1 Attraction periods of some Saturniidae of eastern U.S.

Moth [a]	Attraction Period
Antheraea polyphemus (Cramer)	10:00 PM– 1:00 AM
Actias luna (Linné)	Midnight– 3:00 AM
Samia cynthia (Drury)	6:00 PM–10:00 PM
Hyalophora (*Hyalophora*) *cecropia* (Linné)	3:00 AM– 5:30 AM
Hyalophora (*Callosamia*) *promethea* (Drury)	3:00 PM– 6:00 PM
Hyalophora (*Callosamia*) *angulifera* (Walker)	10:00 PM–11:00 PM

[a] Classification from Michener (1952).

3. POPULATION SAMPLING WITH AN ULTRAVIOLET LIGHT TRAP

For specifications on the design and operation of an ultraviolet light trap see Fox (1963).

A series of some common moth such as the army worm *Pseudaletia unipuncta* (Haworth) should be collected at the light trap and kept alive. These specimens are then marked according to the 1–2–4–7 system (Fig. 6.1) and groups of marked individuals spotted at quarter-mile intervals in a gridded study area about one square mile in size. The ultraviolet light trap should be operated in the center of the area and all marked specimens recovered at the light counted and recorded.

Data are handled in a manner similar to that described for the saturniid moths in the previous experiment. Additional useful information may be gathered by changing the bottle of the light trap at timed intervals during the night. Analysis of the marked individuals recaptured during each time period will give data on the time factor in dispersal and periodicity, if any, of flight.

4. LABORATORY POPULATION CENSUS

Intra- and interspecies population phenomena have been indoor preoccupations of animal ecologists for several decades. Excellent recent reviews on population phenomena are available in Slobodkin (1961) and Macfadyen (1963: Part II, 87–169).

Population census, including age structure, birth, death and reproductive rates have yielded life table data for a number of arthropod species (Birch, 1948; Slobodkin, 1954; Frank *et al.*, 1957; Frank, 1960). A fairly extensive literature has also developed relative to competition in laboratory cultures (Gause and Witt, 1935; Frank, 1957; Huffaker, 1958; Park, 1962). Life table data for natural animal populations were reviewed by Deevey (1947) and a number of ecologists have correlated the information on vital statistics gleaned from the laboratory experiments with field census studies (Edmondson, 1960, 1962; Hall, 1964). Smith (1952) presented a critique of experimental methods in population dynamics; a particularly important point concerns the effects of physiological changes such as acclimation, acclimatization and adaptation (see definitions by Fry, 1958) on the outcome of population experiments.

Studies of population growth and regulation of insect species in laboratory cultures are usually hampered by certain technical problems, not the least of which is the periodic population census itself. As Park (1962) has pointed out, population census procedures should disturb the individuals as little as possible, since following enumeration they are usually returned to culture with the *assumption* that the organisms will act as if they had not been disturbed. Environmental conditions must be strictly controlled if the results are to be repeatable.

If a known number of animals in equal sex ratio of a given species is introduced into an environment in which food and space are not limiting factors and the population is assayed periodically, the general features of population growth for the given environmental conditions can be observed. Naturally, insects with short generation times are more suitable for such laboratory exercises.

MATERIALS. Fruit flies (*Drosophila*) having a generation time of several days to a week at 75° F. or grain beetles, such as *Tribolium confusum* duVal and *T. castaneum* (Herbst) with a generation time of about one month at the same temperature, are well suited for population studies (Park, 1948, 1954, 1962). Since these insects have holometabolous cycles, age distribution, at least in terms of life stages, can be verified. The yellow mealworm *Tenebrio molitor* is a convenient insect to census, but its generation time of several months is a distinct disadvantage. However, since this insect is useful for bioassay in the endocrine studies described in Chapter Four, it can be used for census exercises if cultures are started well in advance so that the student can gather data to the third and fourth generations during the term.

Insects with gradual metamorphosis can also be used—for example, various species of cockroaches or thysanurans—although individual instars and exact reproductive stages are difficult to determine if studies require age structure information.

The confused flour beetle (*Tribolium confusum*) and the red flour beetle (*T. castaneum*) are easily obtained (see Appendix) common grain pests. General information on life history, distribution and structure has been provided in Pfadt (1962). Fine to coarse milled flour, bran or corn meal or mixtures of them constitute suitable food sources for experimental work. The flour used should readily pass through a no. 5, quality XX bolting silk (aperture diameter, 0.28 mm.; the XX means double extra heavy; for data on bolting silks see Welch, 1948) or no. 60 U.S. or Tyler soil testing screens (aperture diameter, 0.25 mm.). Figure 1.7 illustrates the separation of male and female flour beetles by the pattern of strial grooves on the elytra: only in the females does the sixth groove curve in at the apex and unite with the third.

In addition to healthy cultures of *Tribolium*, the following materials are required for both the population studies and the competition experiments (Section A 5): five to ten pound supply of suitable flour, bran or cornmeal (readily passes through 5XX bolting silk); vials or other containers holding 10 to 50 g. of flour; torsion balance; dissecting microscope; sieves made from bolting silk, small sized soil testing screens or bolting silk sieves constructed from coffee can rims to which the silk has been fastened; millimeter rule; forceps with bits of sponge rubber glued at the points of contact to avoid injury to the insects; thermometer; an instrument for measuring humidity, if possible.

Culture cabinets providing temperature and humidity control are most useful for class work, although the six different temperature–humidity regimes (climates) shown in Table 6.2 can usually be set up by using combinations of wet toweling, a desiccant, lamps and a refrigerator. Naturally, precise data can be obtained only with more refined controls of temperature, light and humidity.

Table 6.2 Average numerical abundance of single species populations of *Tribolium confusum* and *T. castaneum* in relation to temperature and humidity (Rank 1 is the most dense, 6 is least dense), and the outcomes of competition between the two species in each of the six different climates [a]

Climate	Rank of *T. confusum* Success when Alone	Rank of *T. castaneum* Success when Alone	Numercial Abundance when Alone Compared	Mixed-Species Competition Outcomes (% of Contests Won)
Hot–Moist	1	3	*conf.* = *cast.*	*conf.* (0), *cast.* (100)
Hot–Arid	6	5	*conf.* > *cast.*	*conf.* (90), *cast.* (10)
Temperate–Moist	2	1	*cast.* > *conf.*	*conf.* (14), *cast.* (86)
Temperate–Arid	3	4	*conf.* > *cast.*	*conf.* (87), *cast.* (13)
Cool–Moist	4	2	*cast.* > *conf.*	*conf.* (71), *cast.* (29)
Cool–Arid	5	6	*conf.* > *cast.*	*conf.* (100), *cast.* (0)

[a] Table modified from Park (1962).

PROCEDURES. The procedure for sampling and enumerating *Tribolium* populations is summarized in Figure 6.2. Two males and two females are introduced into a vial containing a preweighed amount of flour (10–50 g.). A series of such culture vials, appropriately marked, should be kept under each of the climates shown in Table 6.2. The table depicts the expected results when *T. confusum* and *T. castaneum* are cultured separately. Enumeration is accomplished by sieving the cultures every twenty to thirty days. It is desirable to record the numbers of adults, pupae and larvae separately so that the age structure of the various cultures as well as total numbers of individuals can be compared. Eggs are difficult to count because flour adheres to them, but for a complete study they should be included. The eggs retained on the 5XX bolting silk should be carefully reintroduced, along with the other life stages, into fresh flour. The new culture vials should contain the same total weight of flour as did the original vials.

In addition to studies of the effects of temperature and humidity on population growth of flour beetles, the effects of crowding (intraspecies competition) can be studied. This can be readily achieved by decreasing the total volume of food available per individual in the culture. For example, when food supply is short or when animal density is high, cannibalism becomes a factor in the control of population size.

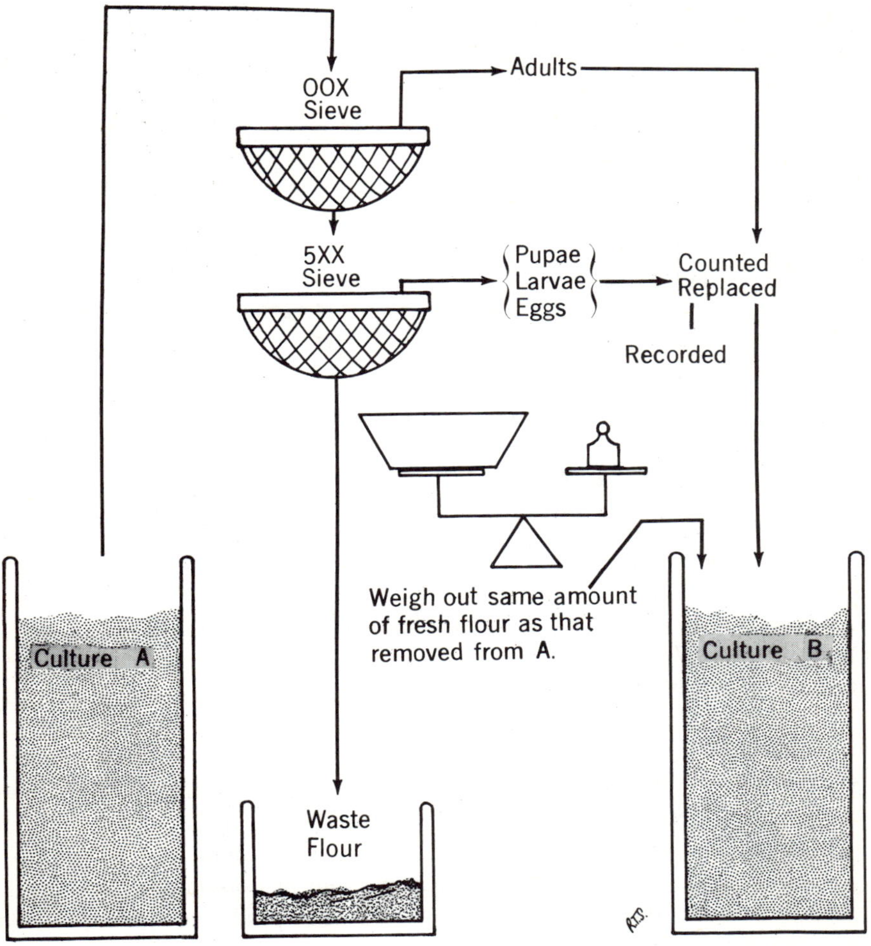

Figure 6.2. Diagrammatic summary of experimental procedure for sampling and enumerating *Tribolium* populations. See text for full explanation.

A set of population census experiments similar to those usually conducted with *Tribolium* has recently been described by Miller (1964) for *Drosophila*. Single species cultures are counted every seven days or less; new cultures are started with first instar larvae removed from the oviposition medium and introduced into vials containing new *Drosophila* food. Miller (1964) counted only the adult flies, weighing them so that the effects of crowding could be evaluated, not only in terms of development time and numbers of adults produced, but also on the basis of adult body weight (or condition). The number of days required for development of *D. melanogaster* and *D. simulans* Sturtevant from early first instar larvae to adults, as influenced by larval density, is shown in Figure 6.3.

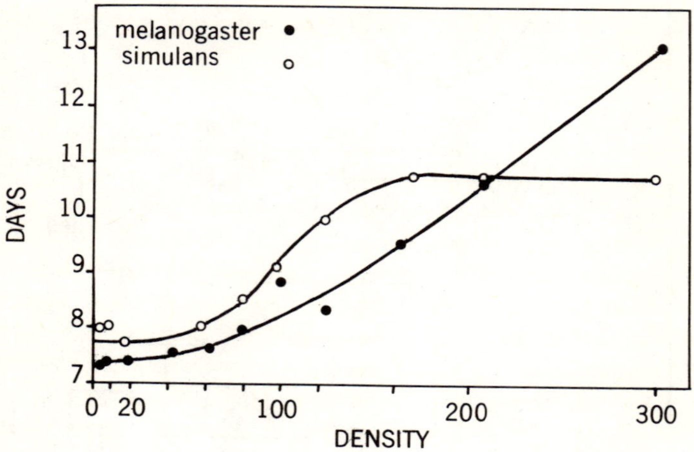

Figure 6.3. Average number of days required for the development of *Drosophila melanogaster* and *D. simulans* from early instar larvae to adults, as influenced by population density. (Modified from Miller, 1964)

Discussion. After the results of the simple experiments described above have been compared with those predicted in Table 6.2, certain additional theoretical applications can be made by advanced class groups.

Considerable refinement in the measurement of population parameters, both in culture and occasionally in the field, has enabled the expression of population phenomena in mathematical terms. The determination of intrinsic rate of natural increase (**biotic potential**) has been emphasized in a number of studies (Lotka, 1945; Birch, 1948; Evans and Smith, 1952; Hall, 1964) and reviewed by Slobodkin (1961) and Macfadyen (1963). In general, the instantaneous rate of population increase can be estimated for a strictly controlled experimental environment—including food supply, temperature, humidity—if the rates of several other population phenomena are known, namely birth, death and, in the field, emigration and immigration. The basic formula is:

$$N_t = N_0 e^{rt}$$

where N_t is the population size at time t and N_0 the initial size; e is the base of natural (Napierian) logarithms, and r is the instantaneous rate of increase. The equation, which assumes that the animals in question are reproducing continuously, is a sufficient approximation if reproduction is a relatively frequent event. Often, as when dealing with laboratory cultures, the effects of death, emigration and immigration may be ignored and birth (b) becomes the instantaneous rate. The equation can then be written (Hall, 1964):

$$N_t = N_0 e^{bt}$$

It is important that either a population having a stable age distribution be utilized or that detailed data be obtained on the way in which the age distribution is continually changing. An important critique of experimental methods in studies of population dynamics has been given by Smith (1952). A considerable discussion has developed over the difference between and, in fact, the definitions of density-dependent and density-independent factors in population control. The theoretical as well as practical implications of this argument are considerable. For a review of alternate theories the reader is referred to Macfadyen (1963).

The details of a trial and error method of successive approximation for calculating r for the rice weevil *Calandra* (*Sitophilus*) *oryzae* is given in Birch (1948). Life table data for *T. confusum* have been given by Park *et al.* (1941).

5. INTERSPECIES COMPETITION

A distinction can be made between two components of the competition process—**exploitation** and **interference**—as discussed by Park (1962). Exploitation operates when the two competing species utilize the same food source, present in a limited supply. Interference refers to interactions between the species such that either reduced reproduction or increased death rate results. For example, high population densities, even though food supply is more than adequate, may result in reduced populations because the overabundant individuals interfere with each other to such an extent that egg laying, or even the act of feeding, are curtailed.

a. Competition Between Tribolium *Species*

Again, *Tribolium confusum* and *castaneum* are ideal subjects for study (Park, 1948; 1954; 1962). Cultures should be started with two pairs of each species and several vials kept under each of the six climate conditions shown in Table 6.2. Population census should be conducted every twenty to thirty days as before, following the procedures shown in Figure 6.2. Once the animals have been sieved from the flour, the adults can be separated by species, utilizing the characters shown in Figure 1.7. The most striking difference involves antennal structure and a hand lens or a dissecting microscope is required for identification and separation.

After two to three months the tabulated results from each temperature–humidity regime can be compared to the expected competitive outcomes shown in Table 6.2. The influence of initial population density on the final competitive result can also be studied by varying the ratios of individuals of each species used to start the experiments.

b. Competition Between Drosophila *Species*

A number of interspecies competition experiments utilizing *Drosophila* have been recorded. All, particularly that of Zimmering (1948), show one fundamental fact: *D. melanogaster* (Meigen) is the most potent competitor and will tend to eliminate other species very rapidly. It is useful, therefore, to inhibit this selective advantage inherent in *melanogaster* by using a deleterious mutant, such as the X chromosome mutant *yellow* (y), recognized by its more or less uniform yellow body coloring.

To facilitate identification, the species used should be readily distinguishable: *melanogaster* in combination with the larger, black-bodied flies of the *obscura* group is ideal. However, members of the *obscura* group have a generation time about double that of *melanogaster,* so that proper timing of the experiment is important. It is suggested that the instructor establish populations a week to ten days before the first laboratory and the experiment should be continued to the end of the term for best results.

Intraspecies competition between genotypes may be attempted as an alternative to interspecies competition; a particularly informative mutant is *Bar eye* (B), an X chromosome gene expressed as a distinctive heterozygote. A plan for such experimentation is given by Strickberger (1962: 47–52), a reference which should be consulted in any case.

The construction of population cages is a problem, but a workable design is given by Strickberger (1962: 48, Fig. 10). Briefly, the cage is made from a standard 13 by 8 by 4½ inch polyethylene pan with a tight fitting cover. A 4 inch square is cut in the lid and replaced by screening. Eight 1⅛ inch holes are drilled or punched in the bottom of the pan, two in the sides and one at the end to receive the food cups. The holes may be fitted with rubber gaskets to hold the cups more securely.

MATERIALS. *Drosophila,* two species or two genotypes, depending on the design of the experiment; at least two population cages; food cups—one ounce cream bottles are best; *Drosophila* food (Appendix); sterilized strips of paper toweling; yeast suspension in dropper bottle; corks of the appropriate size to plug the holes in the population cage; stock culture bottles with *Drosophila* food; plastic spoons; etherizers and ether; constant temperature cabinets.

METHODS. Prepare the food cups by half filling them with *Drosophila* food and store most of them covered. Prepare two bottles per cage as follows: fold a strip of paper toweling and insert it to the bottom through the food so that it serves as a wick; then sprinkle the food with a dropperful of yeast suspension.

The lid should be taped on the cage; fill all the holes except those in the sides. Count out 200 lightly etherized flies of the types to be used, place them in empty food cups and insert into the two side holes of the cage. Write the date on the first two food cups and replace the two corks. The cage is now ready for the experiment.

One of the two population cages should be placed in a temperature cabinet set at 25° C.—the optimum temperature for *melanogaster.* The other should be run at 15° C.—well below the optimum for *melanogaster,* but about ideal for the *obscura* group of species.

Every five days a new food cup should be added. After all the holes are filled, the oldest cup should be replaced. If only *melanogaster* is used, replace food cups twice a week.

Egg samples should be taken every three or four weeks to estimate the frequencies of competing types. To do this put a food cup with a drop of yeast added (no paper) into one of the side holes for a day. Then remove the cup and scrape off the top layer of the food with a plastic spoon; count out 125 to 150 eggs and put them into a fresh culture bottle of food; add yeast and paper. In a few days add more yeast to provide optimum conditions for the larvae. When the adults have hatched, count and identify them as an estimate of frequency.

At the end of the term the flies may be removed from the cage, counted and identified to provide a measure of the reliability of the egg counts.

Throughout this experiment note the changes, if any, of the frequencies of the two species or genotypes and determine which is the more fit (leaves more progeny). Are the trends of selection constant within a cage? Was one species or genotype eliminated? Other pertinent questions are found in Strickberger (1962: 52).

B. COMMUNITY TROPHIC DYNAMICS AND ENERGETICS

The importance of trophic data and the idea of expressing ecological information in energy equivalents are by no means new. As Pearson (1948) reminded, in 1886 Boltzmann (cited in Tizard, 1932) stated that the struggle for existence is essentially a struggle for free energy to do work. Until recently, most data were being gathered and theoretical models constructed by ecologists concerned with aquatic ecosystems

(Shelford, 1913; Naumann, 1925; Thienemann, 1926, 1931; Perfiliev, 1929; Borutsky, 1939; Ivlev, 1939a, 1939b; Juday, 1940; and Lindeman, 1941, 1942).

Since Lindeman's (1942) theoretical statement regarding energy flow in a lake ecosystem, based on the scant data then available, aquatic ecologists have concentrated to a large degree on the measurement of primary productivity—the rate at which plant organisms bind the sun's energy as chemical bond energy (see reviews by Macfadyen, 1948; Strickland, 1960; Doty, 1961).

Satisfactory trophic or food-habit data are available for very few organisms. As Macan (1963) has stated, one of the very first steps in ecological field studies must be to determine the trophic relationships within the community—the **food web.** In cases when the plant or animal host specificity of a given insect species has been determined, the rates at which food is ingested have been measured only rarely (see Banks, 1962; Kloft and Ehrhardt, 1962). In very few instances has the relationship between food ingested and that actually assimilated and used in the organism's metabolism been assessed. The importance of such data is emphasized by studies like that of Smirnov (1962) in which a species of caddis fly larva ingesting two species of vascular hydrophytes obtained suitable nutrition from only one of them. Certainly one of the most important tasks that lies ahead in ecology is the determination of the qualitative and quantitative interrelationships in the food webs of natural ecosystems.

Once the food web has been determined and amounts quantified, the most fruitful approach seems to be the expression of all rate phenomena in one frame of reference—energy equivalents (**calories**). The pathways followed by energy as it enters an ecological system and is degraded at each successive transfer in the food web and the efficiency with which energy is transferred from one food or trophic level to another are of fundamental importance.

Slobodkin (1962: 99) summarized, "It is clear from field data, evolutionary theory, direct calorimetry data and a simple theoretical analysis that energy is of major significance in ecological systems. . . . Even such practical questions as designating optimal procedures for population exploitation and pest eradication are in principle answerable within the existing theoretical framework."

Before trophic dynamic energetic concepts can be put to predictive use, a great deal of data must be gathered. The exercises included in this section are designed to give an insight into some technical and theoretical problems involved in obtaining such data. Whereas most initial trophic studies require little specialized equipment, more detailed analyses do;

this is especially true in the field of energetics. However, its basic importance to the new era of ecology warrants its inclusion; hopefully, many students of the insects soon will be able to begin making calorific determinations on a routine basis.

1. TROPHIC RELATIONS OF INSECTS INGESTING SOLID FOOD

For insects ingesting solid food, qualitative analysis can be accomplished by merely eviscerating the individual under study and identifying the contents of the dissected gut under the compound microscope. Suitable permanent slides can be prepared by using mountants designed for fresh material (see Appendix). The identification of the ingested material is greatly facilitated if slides are also prepared of fragments of possible food substances in the insect's habitat. Some species, such as certain herbivorous insect larvae, are quite specific in their food requirements and analysis is fairly simple.

The quantification of ingested materials can be accomplished conveniently by the following procedure modified from Mecom and Cummins (1964):

1. The ventral surface of the insect is split from the region of the esophagus back to the anus and the entire gut removed to a watch glass containing 10 ml. of distilled water. Either fresh insects or those preserved in a mixture of 70% ethyl alcohol and formalin can be used.

2. The gut is slit with a microscalpel and the contents exposed by gentle pressure on the digestive tract wall; all gut tissue should be removed with fine forceps.

3. The 10 ml. suspension of gut contents is washed into the glass receiving funnel of a microanalysis filter apparatus fitted with a vacuum pump, and filtered. A 25 mm. diameter, gridded Millipore filter with a pore size of 0.45 μ is convenient (see Appendix).

4. The suspension is filtered at very low vacuum; the pump must be shut off and the filter removed while the latter is still slightly wet.

5. One drop of immersion oil is placed on a large microscope slide (38×75 mm.) and the wet filter added on top. An additional two drops of immersion oil are introduced at the center of the filter.

6. The filter is kept covered in a dark place for twenty-four hours, or until the filter has become completely transparent. A 30° C. drying oven hastens the results, but if delicate algae are included in the food, refrigeration and a longer clearing time seem advantageous.

7. After the filter has cleared, add three drops of mountant (Appendix) at the center, put on a cover slip and press out the air bubbles.

8. Dry slides on warming table until they have hardened; it is usually desirable to continue to store the slides horizontally.

9. The slides are counted, using a modification of the procedure given by McNabb (1960). Ten grid squares on the filter surface are selected at random for counting under a compound microscope at 100x, with the ocular fitted with a Whipple disc. The objective is located at random within each of the ten filter grid squares and the kind and number of each food item appearing within the Whipple field of grid squares is recorded. By totaling the counts and expressing the numbers of each food item as a per cent of the total, the relative importance of ingested foods can be shown (Table 6.3).

Table 6.3 Relative % abundance of foodstuffs in guts of selected aquatic insect species in a riffle habitat of Linesville Creek, Pa.[a]

Foodstuffs	*Prosimulium hirtepes* 11 Jan. '64	*Tipula* sp. 11 Jan. '64	*Antocha* sp. 2 Feb. '64	*Chimarra aterrima* 2 Feb. '64
Meridion circulare	63.04	14.40	56.75	71.18
Diatoma vulgare	19.56	6.77	10.13	5.93
Fragilaria capucina	7.60	5.08	0.00	5.08
Nitzachia sp.	1.63	14.40	0.00	0.84
Acnanthes sp.	1.08	0.00	0.00	0.84
Synedra sp.	1.08	0.00	2.02	2.54
Cymbella sp. B	0.54	16.94	0.00	5.08
Cymbella sp. A	0.00	0.00	0.00	6.75
Gomphonema sp.	0.54	10.00	19.59	0.84
Melosira sp.	0.00	6.77	0.00	0.00
Vascular Plant fragments	0.00	7.62	0.00	0.00

[a] From Cummins, Coffman and Roff (1964).

If the entire gut contents of one or more individuals are analyzed following the above procedure, the actual number of food items of each kind per gut can be determined. This is accomplished by relating the filter surface counted to the total effective filter surface. The procedure is given in detail in Publication ADM-10 of the Millipore Filter Corporation.

The procedure for analyzing predator gut contents is the same, except that only head capsules or head parts are counted. Cummins *et al* (1964) counted mandibles of each species; one half of each count then gives the number of individuals of a given species that were ingested. The above authors are working on a key to the mandibles of prey organisms for the stream habitat that is being studied; this greatly facilitates counting predator and omnivore gut contents.

2. TROPHIC RELATIONS OF INSECTS INGESTING FLUIDS

Even though the procedures available for evaluating trophic relations of fluid-ingesting species are still experimental, quite complex and do not lend themselves well to class laboratory work, the entomologist should be aware of this area of investigation. At many institutions it is now possible at least to demonstrate some techniques and provide data for student analysis.

It may well be practical to determine the food species of a fluid-ingesting insect by comparing chromatograms of likely plant or prey species with those made from fluid extracted from the foregut of the herbivore or predator in question. The chromatographic techniques as applied to taxonomic separations by Micks (1954) and Lewallen (1957), discussed previously (Chapter Two) can be utilized.

In many instances the chromatographic technique may not provide sufficient resolution. As pointed out by Macfadyen (1963) the more complex method employed by Dempster (1960; see also Dempster *et al.*, 1960) shows significant promise as a quantitative technique for the investigation of trophic relations. This method is a serological technique in which antisera prepared from prey organisms react with traces of prey-fluid in the predator up to twenty-four hours after ingestion; predator species of the chrysomelid beetle *Phytodecta olivacea* (Forster) were studied. Precipitin test techniques should be particularly useful in studies of fluid-ingesting species and show considerable promise in all studies of trophic dynamics.

3. TRACER TECHNIQUES

Experiments involving radioactive materials *are not recommended for and should not be included in group laboratory programs* at undergraduate level, though the entomology student should be aware of tracer techniques as used for ecological analysis. The discussion following is intended primarily for information only. Advanced or graduate students may find useful ideas and references in the presentation below and individually may be able to duplicate some of the experiments under adequate supervision as orientation to possible original research. The use of radioactive materials requires the utmost caution. Before, during and after experimentation the radiation safety officer (institutional or local governmental) should be consulted.

Radioactive tracer techniques have been employed to advantage in studies concerning insects. Many of these have been reviewed by Jenkins (1957, 1962). In the area of food studies, the trophic relations of aphids have received particular attention (Banks, 1962; Kloft and Ehrhardt, 1962).

Some recent ecologically oriented investigations using tracers have been concerned with measuring ingestion and assimilation of food materials by arthropod species. Notable are the laboratory studies on the microcrustacean *Daphnia pulex* by Richman (1958) and on the mayfly *Stenonema pulchellum* by Trama (1957) and the excellent laboratory-field coordinated study on soil arthropods by Englemann (1961). These and other investigations have been aimed at determining the energy budgets for particular organisms. An important step in all such studies is the separation of ingestion and assimilation rates—determining the difference between the rate of food intake and the rate at which digested food is actually assimilated through the gut wall. The per cent assimilation is often fairly low (Philipson, 1960a, 1960b). The relationship between food intake and utilization are shown by the general formulation

$$\text{ingestion} = \text{assimilation} + \text{egestion}$$

where assimilated energy is apportioned among growth, reproduction and respiration. The last is the maintenance cost for the organism or population, as discussed previously (Chapter Four). In the final analysis each of these categories is best expressed in terms of calories.

The experiments discussed below are conveniently accomplished with carbon-14 or phosphorus-32 as the tracer. Although C^{14} isotope, because its beta emission is of a lower energy, is safer to handle than P^{32}, contamination problems can be severe because of its extremely long half-life (time for one-half the original amount to decay is 5570 years for C^{14}). The higher energy beta emission of P^{32} often makes counting (detecting) the radioactivity in experimental material easier. However, its short half-life (14.3 days) necessitates a decay correction which can be ignored when C^{14} is used.

A gas-flow counter in a lead pig shield, a scaler (amplifier, scaler, register and high voltage supply) and a cylinder of O_2 gas with a regulator (see Appendix) are required for the measurements described below. For the details of counting equipment and procedures one of the many manuals on the subject should be consulted (Arnoff, 1956; University of Michigan School of Public Health, 1957; Picker X-Ray Corporation, 1960).

Radioactivity of the experimental material is measured as counts per given time period as registered on the scaler. These counts are manifestations of the disintegrations of C^{14} (or P^{32}) atoms in the experimental material, which cause ionizations within the detector tube. The absolute radioactivity of a substance can be expressed as disintegrations per second (dps.).Thus a microcurie (μc.) of a radioactive material is defined as that amount having an activity of 3.7×10^4 dps. Since the efficiency of gas–flow detector systems is less than 100%, a measurement of counts per second (cps.) in experimental material must be corrected for this inefficiency—a specific value for each counting system—if the absolute activity in dps. is the value sought. In many experiments, such as those described below, relative counts suffice and the efficiency factor can be ignored. When the radioactivity of a substance is being determined, a correction must be made for the general radiation level in the laboratory where the counts are being made—that is, for background radiation. Since the detector tube is housed within a lead shield (lead pig), the correction for background is typically less than 1 cps.

Radioactive substances, tagged with C^{14} or P^{32}, are available from a number of firms (see Appendix).

Ingestion rate. The determination of ingestion rate of an insect, using C^{14} or P^{32}, can be accomplished by gathering data in the following three categories:

1. Counts per second per unit of food material at time zero, or cps_{T_0}. The food of the insect under study is tagged with C^{14} or P^{32} and the activity (cps.) per unit of food (F) material (cps./g., cps./cell, cps./leaf, cps./prey organism, *etc.*) is determined at the beginning of the experiment (T_0) and at time one (T_{1F}) on remaining food substances. Since the radioactivity per unit of food substance will change at least slightly between time zero and time one, the activity per unit during the experimental period is best taken as an average, or

$$\frac{(cps_{T_0} + cps_{T_1F})}{2}$$

If P^{32} is the tracer utilized, a further correction has to be made for radioactive decay; even an experimental period of several hours will necessitate such a correction. The correction is made by adding the loss due to decay to the value cps_{T_2F}.

Plant foods are tagged by adding radioactive material to nutrient media and many of the compounds used in standard nutrient media are available in tagged form (Appendix). For example, in studies of aquatic insects feeding on algae or vascular hydrophytes, cultures of the plants can be radioactively labeled by the addition of 2 to 10 μc. of C^{14}-tagged sodium bicarbonate or P^{32} as a tagged phosphate (Arnoff, 1956). Terrestrial plants can also be labeled by exposure to an atmosphere containing C^{14}-tagged carbon dioxide; this more complex procedure has been reviewed by Arnoff (1956: Chapter 4).

Activity per unit of algal or detrital food can be determined by collecting two subsamples drawn down on Millipore filters. One subsample is counted in the gas–flow detector system, the other is dried and cleared and the food materials enumerated according to the previously described procedure for trophic analysis.

Vascular plant tissue tagged with C^{14} is best analyzed with a gas-phase counting system (Nuclear–Chicago, 1959, Manual BK-092 and 1961, Manual BK-122; Bernstein and Ballentine, 1950: 1059). The plant tissue to be analyzed for radioactivity is digested by a wet carbon combustion method (Van Slyke *et al.*, 1951) and the C^{14} label collected as $C^{14}O_2$ which is then counted in a gas-phase electrometer system. A significant advantage of gas-phase counting is the 100% detection efficiency, since all the radioactive carbon present in the analyzed material is counted. It is probable that the majority of C^{14} detection studies in ecology will employ this procedure in the future.

If a gas-phase system is not available, vascular plant C^{14} activity can be determined by using a gas-flow detector system if a known amount of tissue is macerated or homogenized prior to plating on planchets. When P^{32} is the tracer utilized, small sections of plant tissue can usually be counted without prior treatment because of the high energy β emissions of this tracer.

The food of bacterial or fungal feeding insects can be labeled by the addition of glucose tagged with C^{14} into growing cultures. This was the procedure employed by Englemann (1961) to label yeast as a food source for the oribatid mites he studied.

If predaceous insects are to be investigated, the prey species can be labeled by first tagging their food supply. Since two transfers of the tag occur prior to measurement of radioactivity in the predator insect, the initial activity levels employed should be higher. As was the case with plant tissue, if P^{32} is the tracer utilized, the initial activity of whole or squashed prey insects can be counted in a gas–flow system without prior treatment. When C^{14} is the tracer, either the gas–phase system should be employed or the prey insects dissolved in hot formamide (Pearce, 1956) and plated on planchets for counting.

2. Counts per second of experimental insects at time zero = 0. The level of activity of the insects to be tested, prior to the initiation of feeding on labeled food, should be at background level—equal to zero in terms of experimentally induced radiation.

3. Counts per second of experimental insects (I) at time one, or $cps_{T_{1I}}$. Activity of the insects that have ingested labeled food can be determined at the end of the experimental period according to the procedures outlined above for prey species in predator studies. The most frequently employed procedure is the hot formamide dissolution. In order to obtain a reasonable estimate of ingestion rate, experiments over several time intervals should be completed. Since $cps_{T_{1I}}$ (and cps_{T_2}, below) will be low due to losses from egestion and respiration, it is important that temperature, humidity and light be controlled during the course of each experiment. The insects (and all other experimental material) should be washed in distilled water prior to treatment for counting.

Since a certain amount of radioactivity will be taken up by the insects through means other than ingestion, a control correction is necessary, cps_{C_1}. Even after a washing treatment with distilled water, a certain amount of radioactive tag remains adsorbed to the insect's body surface. The control insects are kept under the same conditions as the experimental individuals, including exposure to tagged food, but they are prevented from feeding by removal of the mouthparts. Trama (1957) found a greater mortality among his controls, but there seemed to be no effect on noningestion isotope uptake.

The activity of the controls at time one is subtracted from the activity of the experimental individuals to arrive at the actual amount of labeled food ingested per experimental insect

$$cps_{T_{1I}} - cps_{C_1}$$

The procedure and formula for determining ingestion rate are diagrammatically summarized in Figure 6.4.

Assimilation rate. Additional determinations, beyond those made in the ingestion experiments above, are required in order to estimate assimilation rate. A reasonable estimate can be obtained if insects which have ingested radioactive food are held for a postabsorption (P) period prior to counting (Trama, 1957; Englemann, 1961). As in the ingestion experiment, the activity of both experimental (cps_{T_2}) and control individuals (cps_{C_2}) must be determined, following postabsorption, at time two. In this way the nonmetabolic activity still remaining associated with control animals can be subtracted from that of postabsorbed individuals. In order to obtain the best estimates, experimental insects measured at time two should be eviscerated prior to counting.

The per cent assimilation between time one and time two can be determined according to the formula given in Figure 6.4.

Egestion rate. If ingestion and assimilation are known, then egestion can be determined by subtracting assimilation from ingestion. However,

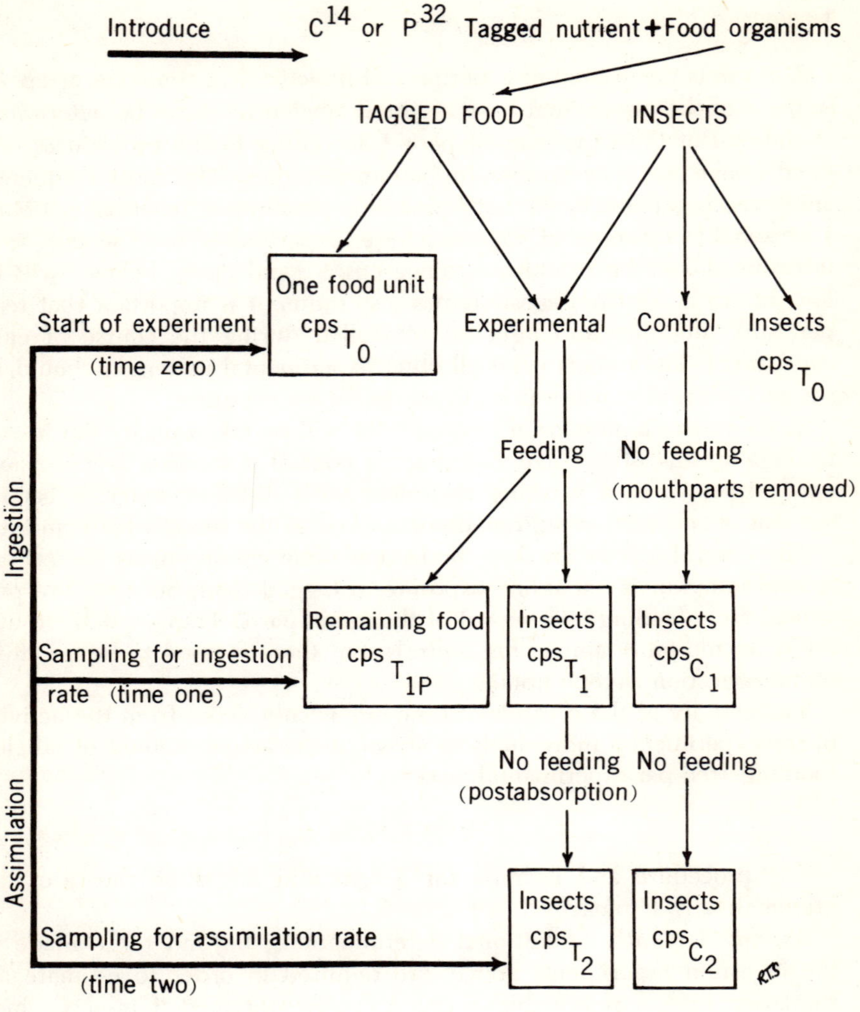

$$\frac{1}{2}\;\frac{cps_{T_{II}} - cps_{C_{I}}}{cps_{T_0} + cps_{T_{IP}}} = \text{Units of food material ingested between time zero and time one}$$

$$\frac{cps_{T_2} - cps_{C_2}}{cps_{T_{II}} - cps_{C_I}} \times 100 = \text{Per cent assimilation between time one and time two}$$

Figure 6.4. Procedural flow and calculations for experimental study of trophic relations. Actual measurements to be made are enclosed in boxes; all cps. values must be corrected for background if background varies. See text for full explanation.

it is important to determine egestion independently whenever possible. In the case of insects that defecate solid material, the feces can usually be collected without too much difficulty. The number of fecal pellets produced between time zero and time one (or between time one and time two) and the activity per unit of fecal material should be measured.

Methods for measuring the rate of egestion by fluid-feeding predators are difficult and limited to use with certain groups such as the Aphidae (Banks, 1962; Kloft, 1962). A method of approximation might be established in which washings of culture containers were collected, evaporated and their radioactivity determined.

4. ENERGETICS

Ecological studies of community energetics have been hampered by the lack of data on energy equivalents (calories) for species under investigation. The conversion of field population census figures to calories requires such information. Some data have begun to appear in the literature (Richman, 1958; Golley, 1961; Comita, 1962; Comita and Schindler, 1963) including some values for certain insect species (Trama, 1957; Golley, 1958; Slobodkin and Richman, 1961).

Study of energy equivalents involves calorimetry and the special apparatus required is expensive. Suitable equipment is available from the Parr Instrument Co. (Appendix), along with detailed instruction manuals (1958, no. 128; 1960, no. 130).

In general, the procedure involves completely burning a sample of known weight under twenty to thirty atmospheres of oxygen. This is accomplished in the special vessel, known as the bomb, so constructed that when it is sealed it will withstand overload pressures up to 200 atmospheres. The heat developed by combustion is absorbed by a carefully determined mass of water in which the bomb is suspended; the water container is known as the calorimeter bucket. The whole system is housed in an insulated cover, the calorimeter jacket. The exact temperature rise in the water is measured with a precision thermometer. From such measurements of temperature rise it is possible to calculate the caloric content of the sample burned, providing a number of corrections are made, including one for heat absorbed by the apparatus rather than the water (the **water equivalent**).

After energy equivalents have been determined for food materials at various life stages of the insect under study—ingestion rate, expenditure of assimilated energy and egestion—it is possible to construct an energy budget for the species in question. An example of such a budget is shown for the mayfly *Stenonema pulchellum* in Table 6.4. As is apparent from the table, data on expenditure of assimilated energy are required before a complete energy budget can be detailed. Rates of growth and reproduction should be assessed, and maintenance cost or respiration expressed as calories (see Chapter Four).

Table 6.4 Energy budget for S. *pulchellum* at 20° C. Based on the mean growth rate, 1.0 mm. increase in length in 33 days [a]

Increase in Length of Nymphs, mm.	Calories Available	Metabolic Loss		Growth		Egestion	
		Calories	Per Cent	Calories	Per Cent	Calories	Per Cent
4.0–5.0	13.53	5.08	37.6	2.48	18.3	5.97	44.1
5.0–6.0	21.02	7.76	37.0	1.98	9.4	11.28	53.6
6.0–7.0	27.56	11.37	41.2	4.32	15.7	11.87	43.1
Mean			38.6		14.5		46.9

[a] From Trama, 1957.

APPENDIX

The purpose of this appendix is to provide information on techniques required for the exercises previously detailed and on sources of supply for certain items not always easily obtained.

Most teaching laboratories have well-established relationships with one or more scientific supply houses. Such firms often are able to provide many things not listed in their regular catalogues or can advise the customer on the nearest or best source of supply. For the information of those using this book and not having a regular supplier for normal items of equipment, some of the firms are:

Aloe Scientific Company, 1031 Olive Street, St. Louis 3, Mo.; Clay-Adams, Inc., 141 East 25th Street, New York 10, N.Y.; Central Scientific Company, 1040 Margin Avenue, Santa Clara, Calif.; Denver Fire Clay Co., 3033 Blake Street, Denver, Colo.; Fisher Scientific Company, 711 Forbes Avenue, Pittsburgh 19, Pa.; LaPine Scientific Company, 2229 McGee Avenue, Berkeley, Calif.; Arthur H. Thomas Company, Vine Street at Third, P.O. Box 779, Philadelphia 5, Pa.

LIVE MATERIALS AND CULTURES

An experimentally oriented laboratory program in entomology requires living insects. It is prudent to establish cultures well before the planned time of use in class. A few of the experiments detailed in the preceding chapters are limited to specific times of the year by the life cycles of the subject species—this is particularly true of field experiments. Most of the laboratory experiments, with planning, can be performed at any time. The subjects for experimentation as specified in each of the exercises are summarized on Table A.1.

There is latitude for substitution, as most of the experiments can be performed using any one of a great variety of subject species. If the species mentioned in the exercises are used, cultures must include *Blaberus* or some other large roach, three genetic strains of *Drosophila melanogaster,* one other Drosophila species, *Tribolium confusum, T. castaneum, Tenebrio molitor* and two or more species of *Gryllus. Anagasta kuhniella* and *Habrobracon juglandis* are needed only for Three C,[1] which can be omitted, but these two insects can

[1] This number and similar numbers throughout the appendix refer to the experiment or exercise discussed in the section cited. *Three C* means the experiment mentioned in Chapter Three, Section C.

Table A.1 Summary of species used in exercises

Exercise	Species Preferred	Condition
Chapter One		
A.1, A.2	*Blaberus,* adults	Alive or preserved
B.1, B.2	Trichoptera, larvae	" " "
C	*Drosophila,* adults	" " "
D.1, D.2, D.3	*Tribolium confusum* and *T. castaneum*	" " "
E.1, E.2	*Danaus plexippus,* adult males and females	Preserved
F.1, F.2, F.3	Collection of insects, myriapods and arachnids	"
Chapter Two		
A.1	Collection of insects, myriapods and arachnids	"
B.1.a	Collection of Trichoptera larvae with cases	Alive or preserved
B.1.b	Collection of *Gryllus* species	Alive
B.1.c, B.2	Two or more species of *Drosophila*	"
Chapter Three		
A.1	*D. melanogaster,* wild and vestigial wing strains	Cultures
A.2	*D. melanogaster,* wild and Bar eye strains	"
B	*Drosophila,* larvae	Alive
C	*Habrobracon juglandis* and *Anagasta kuhniella*	Cultures
Chapter Four		
A.1	Saturniidae, adult males and females	Alive
A.2	Flies: *Sarcophaga, Phormia, Musca* or *Drosophila*	"
A.3.a	*Danaus plexippus,* larvae	"
A.3.b	*Papilio asterius,* larvae	"
A.3.c	*D. plexippus* and *P. asterius,* larvae	"
A.4, A.5	*Tribolium,* mosquitoes or midges	"
B.1, B.2, C.1	*Blaberus* or other insect, arachnid or myriapod	"
C.2, C.3	Trichoptera larvae or other aquatic larvae	"
D.1, D.2, D.3	*Blaberus* or other insect, arachnid or myriapod	"

Table A.1 Summary of species used in exercises (cont.)

Exercise	Species Preferred	Condition
Chapter Four		
E.1, E.2	*Blaberus*	Alive
F.1	*Hyalophora cecropia,* adults	"
F.2, F.3	*Tenebrio molitor,* pupae	"
Chapter Five		
A.1	Trichoptera, larvae, preferably Phryganeidae	"
A.2	Trichoptera, larvae, preferably Limnephilidae	"
B	*Drosophila*	Cultures
C	Same *Gryllus* species used in Two B.1.b	Alive
D	Spiders, preferably Araneidae, with webs	"
Chapter Six		
A.1	*Colias eurytheme* or other common butterfly (field)	"
A.2	*Hyalophora cecropia* or other Saturniidae	"
A.3	Any locally common moth (field)	"
A.4, A.5	Two species each of *Drosophila* and *Tribolium*	Cultures
B.1	Any species ingesting solid food	Alive
B.2, B.3, B.4	Demonstrations or discussions	

be useful alternatives or additional subjects for some of the studies in physiology. The menagerie should also include *Hyalophora cecropia* and other Saturniidae, some spiders and some caddis flies—all alive but not necessarily as breeding cultures.

One or both of the following references should be at hand: Galtsoff *et al.* (1937); Peterson (1934–1937). The former details culture methods for invertebrates in general and includes information on a large number of insect species. The latter is in two volumes and deals with equipment, including cages and traps, in Part I (1934) and abstracts a large number of published reports on breeding, culture methods and special techniques in entomology in Part II (1937).

A wide range of insects, arachnids and myriapods—the species and numbers vary from year to year—are available from supply houses specializing in live material. Three large firms with generally good selection are:

Carolina Biological Supply Company, Burlington, N.C.; General Biological

Supply House, Inc. (Turtox), 8200 South Hoyne Avenue, Chicago 20, Ill.; Ward's Natural Science Establishment, Inc., P.O. Box 1712, Rochester, N.Y. 14603, and (Ward's of California) P.O. Box 1749, Monterey, Calif. 93942.

Local sources should not be overlooked. Dealers in aquarium supplies or in fish bait often can provide cultures of suitable species. Live crickets, for example, are commonly used as bait.

The best local source of all is, of course, the natural habitats of desired species. In a particular region or at a particular time of the year it may be best to establish cultures of local species and to use them as substitutes in some of the exercises.

Whatever the source of the material, the care and maintenance of living specimens and cultures can be made a useful and instructive project for one or more students.

BUTTERFLIES

Larvae of the monarch butterfly, *Danaus plexippus,* feed on milkweeds. They may be obtained by collecting them in the field at the suitable time of the year or by confining females in net cages with the growing plant. Preserved adults are offered by some dealers and are suitable for dissection.

The common black eastern swallowtail, *Papilio asterius,* and its close relatives in the Rocky Mountain and Pacific Coast regions feed on umbelliferous plants. Planting a row of parsley in the back yard is usually sufficient to provide a supply of caterpillars. Although we have not attempted to do so, colonizing this butterfly in large breeding cages would seem to be worth trying.

Insects in general are far more tolerant of cold than of high heat. Butterfly larvae may be retarded in their development without harm by being kept in a cool, moist environment and thus can be used in the laboratory long after the time when they can be found in the field.

COCKROACHES

Blaberus and other roaches with similar habits are easily kept in culture in battery jars or similar moderately large containers. Covers should be tight and consist of an inner layer of gauze kept in place with rubber bands and an outer layer of screen wire or hardware cloth to prevent depredations from stray mice or rats. A small roll of screen wire should be placed in the jar to afford a hiding place for the roaches. Water is provided in shell vials with cotton wicks and plugs. Any of the standard laboratory rations for mice or rats, such as Purina Lab Chow, can be used for feeding but should be supplemented from time to time with some slices of apple. A reasonably warm and humid environment should be maintained.

DROSOPHILA

Stock cultures of wild type *Drosophila melanogaster* and of various mutant strains are available from most supply houses dealing with living material. It may be more difficult to obtain other species. One source is Curator of *Drosophila* Stocks, Rockefeller Institute, New York, N.Y. 10021. Other sources are listed in volumes 35 and 37 of *Drosophila Information Service,* a journal available from Department of Biology, University of Oregon, Eugene, Ore.

Local species may be obtained by trapping.

Many formulations for *Drosophila* food have been used. The simplest for a limited number of cultures is a cornmeal-molasses-agar mixture. Thoroughly wet 10 g. cornmeal with 20 cc. water. Dissolve 1.5 g. agar in 30 cc. water, bring to a boil and add 15 cc. water, stirring to hasten solution of the agar. Add the cornmeal-water suspension and 13.5 cc. molasses to the agar water. Boil the entire mixture for five to ten minutes. Add 10 cc. water and a small amount of fungicide (available for the purpose from most supply houses) and pour the mixture into pint or half-pint milk bottles, filling each only one-third to one-half. Plug bottles with cotton and refrigerate until ready to use. Just before using a bottle, shove a strip of sterilized paper towel into the medium and wet it with a drop or two of aqueous suspension of brewer's yeast. Other formulae are given by Galtsoff *et al.* (1937) and Strickberger (1962).

GRYLLUS

Cover the bottom of a battery jar or equivalent container with about one inch of sand, which should be kept slightly moist at all times. A cover is needed if the jar is shallower than eight inches, but taller jars are more desirable. Screen wire or hardware cloth covers should be used if mice or rats are present. These crickets may be fed on rolled oats only. A better ration is rolled oats ground with some sugar and powdered milk and moistened with water; the mixture is spread thickly on heavy paper and dried. A small section of paper one or two inches square will sustain several crickets if replaced every two or three days. A water supply is not necessary if the sand is kept moist.

HABROBRACON AND ANAGASTA

Culturing this parasitic wasp requires cultures of the moth. Martin (1947) details the life histories of both insects and gives many refinements in handling them for genetic work.

Shallow cardboard suit boxes used by tailors and department stores are ideal for the moth cultures. A layer of yellow cornmeal, rolled wheat or oatmeal is scattered over the bottom and about ten males and ten females are introduced. The eggs, dropped at random over the surface of the food, hatch in about a week. After two or three weeks an inspection should reveal webbiness in the cereal where young larvae have attached particles of food with silk. Temperatures between 27° and 30° C. and humidity between 80 and 90 per cent give best results for development. Adults appear about six weeks after the eggs are laid. Development of larvae may be arrested by placing the culture in a cold place, thus prolonging their availability for use with the wasp. For extensive work with *Habrobracon,* it is best to set new moth cultures every week so that last instar larvae of *Anagasta* will be always at hand.

Males and females of the wasp may be kept alive in shell vials for extensive periods by feeding on a mixture of honey and water. At ordinary room temperature it is necessary to feed the wasps every other day. If it is desired to keep individuals isolated for extended times (up to months) they may be fed once and placed in a refrigerator at approximately 10° C.

Females are given one to five moth larvae in a shell vial. After a sufficient number of eggs have been deposited in the prey, the female may be removed and saved. If the shell vial with the stung larvae is placed on its side, the wasps will pupate separately on the walls of the vial; at this point the residual moth carcasses should be removed. Pupation lasts for three days; on the third day the individuals may be sexed by observing the well-developed antennae and genitalia and virgin females can be isolated before eclosion.

Habrobracon stocks and *Anagasta* cultures are available from many dealers in live insects.

SATURNIIDAE

Cocoons of various of the giant silk worm moths have been offered sporadically for many years by individuals and sometimes are advertised in journals catering to nature hobbyists. The increasing importance of *Hyalophora cecropia* in the laboratory has led to this species being listed at times by regular commercial houses but demand generally exceeds supply.

Nearly mature larvae and cocoons are best collected locally in season. With them or with adults attracted to light traps it is possible to establish most Saturniidae as cultures in large cages if a ready supply of the correct plant is at hand. For food plants see Collins and Weast (1961) or Michener (1952). The former reference also gives suggestions on rearing various species and information on the seasonal correlation of the life cycle in the United States.

Development of larvae may be arrested so that they can be saved for later laboratory use by keeping them in a cool, somewhat moist environment. Adult eclosion often can be hastened by chilling the cocoon for a few days, then bringing it into a warm environment.

SPIDERS

Lougee's (1963) method of collecting suspended spider webs is summarized in Five D. Spiders may be kept alive for considerable periods of time in terraria or large jars if the natural habitat is reasonably approximated. Both *Drosophila* and *Anagasta* are suitable spider food. Adults can be released into the covered terrarium. Hunting spiders should be provided with a suitable hiding place and the bottom of the container should be covered with sand, pebbles, leaves, moss or humus as appropriate to the species. Nymphs of cockroaches or grasshoppers can be used to feed them. A shell vial of water-soaked cotton should be placed in the cage.

TENEBRIO

Any shallow container may be used and does not need to be large. Galvanized iron boxes two feet by one foot by one foot deep have been recommended, but battery jars, aquaria or gallon jars will also serve. Cover the bottom to a depth of about one-quarter inch with bran mash or chick mash, then with four or five layers of burlap with a little mash between each layer. Introduce up to hundreds of larvae, pupae and/or adults. Sprinkle the top layer of burlap with water daily. Avoid excess moisture and mold. About 30° C. is the optimum temperature.

TRIBOLIUM

Put about 200 g. whole wheat or patent flour into a pint milk bottle and introduce about ten males and ten females; stopper with cotton. It is desirable to maintain temperature at 28° to 30° C. and to keep humidity between 25 and 75 per cent. Flour in the culture bottles should be changed periodically. After about twenty days pupae may be removed by sifting them out of the flour and they may be sexed by examination under a dissecting microscope. Female pupae bear a pair of small appendages at the tip of the abdomen; male pupae have these appendages very much reduced. Select the necessary number of each sex for the new cultures.

TRICHOPTERA LARVAE

Representatives of the caddis fly families Phryganeidae and Limnephilidae are generally common everywhere in unpolluted streams. They should be hunted in vegetation and organic debris along stream margins. Shallow parts of lakes and ponds among stands of aquatic vegetation also are good collecting locations. Species of the genera *Limnephilus* or *Platycentropus* often are present in large numbers in temporary ponds.

Larvae can be kept alive in dishes and aquaria by duplicating as well as possible the conditions of the natural habitat. Species from standing water are easier to maintain because no special equipment such as pumps and aerators is required.

MEDIA AND FORMULATION

Most of the media, reagents and stains mentioned in the exercises are commonly available from any supply house. For the rest, notes are supplied below.

JUVENILE HORMONE

Many juvenile hormone analogues are expensive and difficult to obtain. A commercial grade of farnesol, however, is available from General Biochemical Company, Shagrin Falls, Ohio. Farnesyl methyl ether and farnesyl diethylamine may be obtained from Hoffmann La Roche, Inc., Basel, Switzerland.

LACTO-ACETO-ORCEIN SOLUTION

Heat 35 cc. glacial acetic acid to boiling and add 1 g. of orcein. Boil the mixture for twenty minutes and add 35 g. lactic acid; continue to boil for another ten minutes. Cool to room temperature and filter; add 30 cc. water before using.

MOUNTANTS, PERMANENT

Canada balsam, the classic mountant, has been almost universally supplanted by synthetic mountants because it remains fluid or tacky for so long a time, allowing the objects mounted to drift about, and because it becomes discolored and more or less opaque. For whole mounts especially, the synthetic media, which dry in a few hours to a few days and remain translucent permanently, are superior.

Euparal is available from various firms, including General Biological Supply House, Chicago, Ill. Objects may be mounted directly from alcohol or may first be passed through xylene, as preferred.

HSR (Harleco Synthetic Resin) is available from Hartman-Leddon Company, 60th and Woodland Avenue, Philadelphia, Pa. 19143 (Catalogue 7885, ¼ pound). The crystals supplied are soluble in xylene or toluene and objects must be passed through the solvent used before mounting.

Permount is available from Fisher Scientific Company, 711 Forbes Ave., Pittsburgh 19, Pa. as a thin solution in toluene designed primarily for histological preparations. It can be adapted for use in whole mounting by evaporating it to the desired viscosity.

MOUNTANTS, TEMPORARY

Living specimens small enough to make into whole mounts may be placed directly into one of the mountants listed below. A general review of mounting living insects for study is given by Frings (1946). Mounts made with these media must be regarded as essentially temporary; their advantage lies in ease of preparation where a permanent mount is not wanted and sometimes in favorable indices of refraction for special study. The useful life of such a preparation can be prolonged if the cover slip is sealed to the slide with a suitable material as given in any reference on microscopy.

CMC-10 and CMC-S are sold by General Biological Supply House, Chicago, Ill.

PVL (polyvinyl lactophenol) is available from George T. Gurr, Ltd., London, S.W. 6, England.

Andre's medium is detailed and discussed by Gottlieb (1963).

NINHYDRIN SPRAY

Ninhydrin spray is supplied by Sigma Chemicals, St. Louis 18, Mo.

PRINGLE'S SALINE SOLUTION

To 1 liter of 0.05*M* phosphate buffer (pH 6.5) add 9.0 g. NaCl, 0.2 g. KCl, 0.2 g. $CaCl_2$ and 4.0 g. glucose.

SARKARIG'S STAIN

To make 100 ml.: dissolve 0.5 g. crystal violet in 95 ml. distilled water and add 1.090 g. NaCl, 0.157 g. KCl, 0.085 g. $CaCl_2$ and 0.157 g. $MgCl_2$. When the salts are completely dissolved, acidify with acetic acid (0.125 ml. glacial acetic acid) and bring the total volume to 100 ml. with distilled water; filter. The solution is stable for about thirty days.

TAP WATER CONDITIONING

Four C.2 on oxygen consumption by aquatic insects emphasizes the importance of conditioning tap water if it is used in such an experiment. Tap water should be aerated vigorously for at least twenty-four hours. Alternatively, it may be filtered slowly through a charcoal column.

WHOLE MOUNTS

All manner of complicated procedures have been recommended for making whole mounts and genitalic slides. It is generally true in microscopy, however, that the best procedure is the simplest one which yields the results desired. If a whole mount is needed only for temporary examination and study and a permanent mount is not wanted, any of the media listed above as temporary mountants can be used. As noted, preparations with such media can be made semipermanent or even permanent, but always are less satisfactory if only because the objects tend to drift.

For a permanent whole mount, one of the synthetic mountants should be used and of these we have found HSR to be most convenient and satisfactory. The crystals supplied are soluble in xylene or toluene and the solution can be adjusted to the desired viscosity. Two viscosities should should be on hand, a thin watery solution and a thick molasses-like solution.

Genitalia or other dissected sclerotized parts are passed through 95 per cent alcohol—absolute alcohol is not necessary, nor is the alcohol ladder advocated by some technicians—directly from water, then through xylene or toluene. Time required for each step depends on the size of the object and varies from ten to thirty minutes. Extra time is safe and does no harm; in fact, a full day in either reagent would not damage the objects.

Whole insects, mites, ticks and the like should first be punctured with a fine needle to permit thorough penetration by the reagents; the best location for puncturing is wherever the punctures will be the least noticed and will least obscure structures—usually this is along the pleural region on one side of the abdomen. Sometimes it is also desirable to force out the contents of the gut. Specimens are then passed through 95 per cent alcohol and xylene and generally require at least one or two hours for each step.

To mount, place a drop of thin mountant on the slide and orient the object or objects. Three tiny bits of glass or plastic sheeting of appropriate thickness to prevent undue flattening of the specimen when covered should be arranged around the object. The slide is then set aside for a few hours or overnight, allowing the mountant to become tacky or hard. Then cover the object with a drop of the molasses-like mountant, wet the underside of the cover slip with xylene and cover. This two-step procedure in mounting ensures that the object mounted will maintain the desired orientation and is particularly useful for genitalia with parts dissected.

For dissection, puncturing and evacuating gut contents the standard 67 mm. syracuse watch glass is convenient. A little alcohol added to water reduces the meniscus and makes dissection easier. For the alcohol and xylene steps, syracuse watch glasses can be used but time is saved by using the smaller 27 mm. U.S. Bureau of Plant Industry watch glass (Arthur H. Thomas Co., P.O. Box 779, Philadelphia 5, Pa. catalogue 9850).

SPECIAL EQUIPMENT

Newman (1964) is an invaluable reference for general biological instrumentation. Most equipment necessary can be purchased from any of the regular supply houses. Many of these firms publish excellent manuals pertaining to special aspects of instrumentation; catalogues should be consulted for the lists of those available.

HUMIDITY AND TEMPERATURE MEASURING DEVICES

Hygrometers and thermometers for various special purposes and meeting various standards of sensitivity and accuracy are listed in the catalogues of most supply houses. Instruments especially designed for biological uses are also available from:

Sardex, Inc., 12 Bowdoin Square, Boston 14, Mass. (hygrometer); Applied Research Associates of Texas, Inc., 6541 North Lamar Boulevard, Austin 5, Tex.; Rustrak Instrument Company, Inc., 130 Silver Street, Manchester, N.H.; Tri-R Instruments, 144-13 Jamaica Avenue, Jamaica 35, N.Y.

INSECT SONGS

A recording of insect songs is available from General Biological Supply Co., Chicago, Ill. (catalogue 130T507) and is entitled, "Calls of the common crickets, grasshoppers, cicadas and other insects of the eastern United States."

MILLIPORE FILTERS

Apparatus and information are available from Millipore Filter Corporation, Bedford, Mass.

NERVE CHAMBER

A nerve chamber suitable for use with insects can be obtained from Harvard Apparatus Company, Dover, Mass.

OXYGEN BOMB CALORIMETRY

Instrumentation and operation manuals are available from Parr Instrument Company, Moline, Ill.

OXYGEN ELECTRODE INSTRUMENTS

The specialized oxygen electrode instruments used for measuring dissolved oxygen concentration are available from:

Beckman Instruments, Inc., Scientific and Process Instruments Division, 2500 Harbor Boulevard, Fullerton, Calif.; Chemtronics, Inc., P.O. Box 417, San Antonio, Tex. (gas phase analyzer); Precision Scientific Company, 3737 West Courland Street, Chicago 47, Ill.; Yellow Springs Instrument Company, Inc., Yellow Springs, Ohio.

TRACER EQUIPMENT

Radiation detection and counting equipment is offered by:

Nuclear-Chicago Corporation, 333 East Howard Avenue, Des Plaines, Ill.; Picker X-Ray Corporation, White Plains, N.Y.; Planchets, Inc., Chelsea, Mich.; Victoreen Instrument Company, 5806 Hough Avenue, Cleveland 3, Ohio.

Tagged compounds such as C^{14} and P^{32} may be obtained from:

Oak Ridge National Laboratory, Radioisotopes: Special Materials and Services, Oak Ridge, Tenn.; Nichem, Inc., P.O. Box 5737, Bethesda 14, Md.; Radiochemical Division, Nuclear-Chicago Corporation (above).

KEYS

There is little need for students in an introductory course in entomology to identify specimens to genus and species. Those who wish to do so must possess a knowledge of structure and a vocabulary beyond that which may be expected of a beginning student and should be referred to the voluminous monographic literature. Fox and Fox (1964: 411–412) provide a list of selected references to classes and orders and many of these papers contain keys to the groups covered.

For use in a general course, keys need only carry identification at most to the family. The little, illustrated brochure by Metcalf and Metcalf (1928) includes the insect families commonly encountered and is very satisfactory for class use. It is available at $1.50 each from Mrs. K. Metcalf Browne, 2728 Cambridge Road, Raleigh, N.C. or from Mrs. Cleo F. Metcalf, 1757 Prince Albert Drive, Riverside, Calif.

In the "How to know" series, Jacques (1947) has keys to adult insects and Chu (1949) to immature insects. Both books are illustrated and are useful for North American families. Borrer and De Long (1964) has keys to most groups found in the United States. Encyclopedic treatment of the families of Insecta, Myriapoda and Arachnida is offered by Brues, Melander and Carpenter (1954).

REFERENCES CITED

Alexander, R.D., 1957a. The taxonomy of the field crickets of the eastern United States (Orthoptera: Gryllidae). Ann. Amer. Ent. Soc., 50: 584–602.

Alexander, R.D., 1957b. Sound production and associated behavior in insects. Ohio Jour. Sci., 57: 101–113.

Alexander, R.D., and R.S. Bigelow, 1960. Allochronic speciation in field crickets, and a new species, *Acheta veletis*. Evolution, 14: 334–346.

Alexander, R.D., and T.E. Moore, 1958. Studies on the acoustical behavior of seventeen-year cicadas (Homoptera: Cicadidae). Ohio Jour. Sci., 58: 107–127.

Alexander, R.D., and T.J. Walker, 1962. Two introduced field crickets new to eastern United States (Orthoptera: Gryllidae). Ann. Ent. Soc. Amer., 55: 90–94.

Allee, W.C., and R. Oesting, 1934. A critical examination of Winkler's method for determining dissolved oxygen in respiratory studies with aquatic animals. Physiol. Zool., 7: 509.

American Public Health Association, Inc., 1960. Standard methods for the examination of water and wastewater including bottom sediments and sludges (11th ed.) Amer. Pub. Health Assoc., Inc., 1790 Broadway, New York, 626 pp.

Arnoff, S., 1956. Techniques of radiobiochemistry. Iowa State Univ. Press, Ames, Iowa, 228 pp.

Baldwin, E., 1948. An introduction to comparative biochemistry (3rd ed.) Cambridge Univ. Press, Cambridge, England, 164 pp., ill.

Banks, C.J., 1962. Some recent studies involving the use of radioisotopes . . . feeding behavior of two phytophagous insects. *In* Proc. of symposium on radioisotopes and radiation in entomology: 175–179. International Atomic Energy Agency, Vienna.

Bartels, M., and F. Baltzer, 1928. Über Orientung und Gedächtnis der Netzspinne, *Agalena labrinthica*. Rev. Suisse Zool., 35: 247–258.

Basrur, V.R., and K.H. Rothfels, 1959. Triploidy in natural populations of the black fly *Cnephia mutata* (Malloch). Canadian Jour. Zool. (Ottawa), 37: 571–589.

Bastock, M., and A. Manning, 1955. The courtship of *Drosophila melanogaster*. Behaviour, 8: 85–111.

Bates, M., 1949. The natural history of mosquitoes. Macmillan, New York, 379 pp., ill.

Berg, K., 1952. On the oxygen consumption of Ancylidae (Gastropoda) from an ecological point of view. Hydrobiologia, 4: 225.

Berg, K., 1953. The problem of respiratory acclimatization illustrated by experiments with *Ancylus fluviatilis* (Gastropoda). Hydrobiologia, 5: 331–350.

Berg, K., J. Lumbye and K.W. Ockelmann, 1958. Seasonal and experimental variations of the oxygen consumption of the limpet *Ancylus fluviatilis* (O.F. Müller). Jour. Exp. Biol., 35: 43–73.

Berg, K., and K.W. Ockelmann, 1959. The respiratory requirements of freshwater snails. Jour. Exp. Biol., 36: 690–708.

Bernstein, W., and R. Ballentine, 1950. Gas phase counting of low energy β-emitter. Rev. Sci. Instruments, 21: 158–162.

Betten, C., 1934. The caddis flies or Trichoptera of New York State. Bull. New York State Mus., no. 292: 1–576.

Birch, L.C., 1948. The intrinsic rate of natural increase of an insect population. Jour. Animal Ecol., 17: 15–26.

Bliss, D.E., and D.M. Skinner, 1963. Tissue respiration in invertebrates. Amer. Mus. Nat. Hist., New York, 139 pp.

Block, R.J., E.L. Durrum and G. Zweig, 1958. A manual of paper chromatography and electrophoresis. Academic Press, New York, 710 pp., ill.

Bodenstein, D., 1953. The role of hormones in molting and metamorphosis. *In* Roeder, 1953: 879–932.

Borror, D.J., and D.M. De Long, 1964. An introduction to the study of insects (2nd ed.). Holt, Rinehart & Winston, New York, 819 pp., ill.

Borror, D.J., and C.R. Reese, 1953. The analysis of bird songs by means of a vibrilizer. Wilson Bull., 65: 271–276.

Borutsky, E.V., 1939. Dynamics of the total benthic biomass in profundal of Lake Beloie. Proc. Kossino Limnology Stat. Hydromet. Serv. U.S.S.R., 22: 196–218. Translated by M. Ovchynnyk, edited by R.C. Ball and F.F. Hooper, reproduced by Michigan Dept. Conserv.

Brown, R.G.B., [1964]. A comparative study of mating behaviour in the *Drosophila obscura* group. Animal behaviour, in press. Oxford Univ., England, Ph.D. thesis, 1962.

Brues, C.T., A.L. Melander and F.M. Carpenter, 1954. Classification of insects. Bull. Mus. Comp. Zool., 108: 1–917.

Buck, J.B., 1953. Physical properties and chemical composition of insect blood. *In* Roeder, 1953: 147–190.

Buck, J.B., 1962. Insect respiration. Ann. Rev. Ent., 7: 27–56.

Buck, J.B., and M. Keister, 1955. Cyclic CO_2 release in diapausing *Agapema* pupae. Biol. Bull., 109: 144–163.

Bullough, W.S., 1951. Practical invertebrate anatomy. Macmillan, New York and London, 463 pp., ill.

Buzzati-Traverso, A.A., 1960. Paper chromatography in relation to genetics and taxonomy. *In* Walker, P.M.B. (ed.), 1960, New approaches in cell biology: 95–123. Academic Press, New York.

Buzzati-Traverso, A.A., and A.B. Rechnitzer, 1952. Paper partition chromatography in taxonomic studies. Science, 117: 58–59.

Campbell, F.L. (ed.), 1959. Physiology of insect development. Univ. Chicago Press, Chicago, 167 pp.

Carthy, J.D., 1958. An introduction to the behaviour of invertebrates. G. Allen, London, 380 pp., ill.

Cassidy, H.G., 1957. Fundamentals of chromatography. Interscience, New York, 447 pp.

Chu, H.F., 1949. How to know the immature insects. Wm. C. Brown Co., Dubuque, Iowa, 234 pp., ill.

Collins, M.M., and R.D. Weast, 1961. Wild silk moths of the United States. Collins Radio Co., Cedar Rapids, Iowa, 138 pp., ill.

Comita, G.W., 1962. The energy budget for *Diaptomus siciloides* Lilljeborg. Verhandl. Internatl. Ver. Limnologie, 15: 646–653.

Comita, G.W., and D.W. Schindler, 1963. Calorific values of microcrustacea. Science, 190: 1394–1396.

Comstock, J.H., 1940. An introduction to entomology. Cornell Univ. Press, Ithaca, N.Y., 1064 pp.

Comstock, J.H., 1948. The spider book; a manual for the study of the spiders and their near relatives. Rev. and ed. by W.J. Gertsch. Cornell Univ. Press, Ithaca, N.Y., 729 pp., ill.

Cragg, J.B. (ed.), 1962. Advances in ecological research. vol. 1. Academic Press, New York, 203 pp.

Cummins, K.W., [1961]. The microdistribution of the caddis fly larvae *Pycnopsyche lepida* (Hagen) and *Pycnopsyche guttifer* (Walker) in a restricted portion of a small Michigan stream. Univ. Michigan, Ph.D. dissertation, 158 pp.

Cummins, K.W., 1962. An evaluation of some techniques for the collection and analysis of benthic samples with special emphasis on lotic waters. Amer. Midland Nat., 67: 477–504.

Cummins, K.W., [1964]. Factors limiting the microdistribution of the caddis flies *Pycnopsyche lepida* (Hagen) and *Pycnopsyche guttifer* (Walker) in a Michigan stream (Trichoptera: Limnephilidae). Ecol. Monogr. [In press.]

Cummins, K.W., W.P. Coffman and P.A. Roff, [1964]. The food web of Linesville Creek, Pennsylvania. [In preparation.]

Dam, L. van, 1935. A method for determining the amount of oxygen dissolved in 1 cc. of water. Jour. Exp. Biol., 12: 80–85.

Dawson, R.M.C., D.C. Elliott, W.H. Elliott and K.M. Jones (eds.), 1959. Data for biochemical research. Oxford Univ. Press, London, 299 pp.

Deevey, E.S., Jr., 1947. Life tables for natural populations of animals. Quart. Rev. Biol., 22: 283–314.

Demerec, M., and B.P. Kaufmann, 1962. Drosophila guide. Carnegie Inst. Washington, 47 pp., ill.

Demster, J.P., 1960. A quantitative study of the predators on the eggs and larvae of the broom beetle, *Phytodecta olivacea* Forster, using the precipitin test. Jour. Animal Ecol., 29: 149–168.

Demster, J.P., O.W. Richards and N. Waloff, 1960. Carabidae as predators on the pupal stage of the chrysomelid beetle *Phytodecta olivacea* Forster. Oikos, 10: 65–70.

Denning, D.G., 1956. Trichoptera. *In* Usinger, 1956: 237–270.

Dethier, V.G., 1937. Gustation and olfaction in lepidopterous larvae. Biol. Bull., 72: 7–23.

Dethier, V.G., 1941. Chemical factors determining the choice of food plants by *Papilio* larvae. Amer. Nat., 75: 61–73.

Dethier, V.G., 1947. Chemical insect attractants and repellants. Blakiston, Philadelphia, 289 pp.

Dethier, V.G., 1953. *In* Roeder, 1953. (a) Vision, 488–522. (b) Mechanoreception, 523–543. (c) Chemoreception, 544–576.

Dethier, V.G., 1956. Chemoreceptor mechanisms. *In* Grenell, R.G., and L.J. Mullins (eds.), 1956: 1–30, Molecular structure and functional activity of nerve cells. Publ. Amer. Inst. Biol. Sci.

Dethier, V.G., 1962. To know a fly. Holden-Day, San Francisco, 119 pp., ill.

Dethier, V.G., and L.E. Chadwick, 1948. Chemoreception in insects. Physiol. Rev., 28: 220–254.

Dill, D.B., E.F. Adolph and C.G. Wilber (eds.), 1964. Adaptation to the environment. Handbook of physiology, section 4. Amer. Physiol. Soc., Washington, D.C., 1056 pp.

Dixon, M., 1951. Manometric methods as applied to the measurement of cell respiration and other processes. Cambridge Univ. Press, London, 165 pp., ill.

Dobzhansky, T.G., 1951. Genetics and the origin of species (3rd ed., rev.). Columbia Univ. Press, New York, 364 pp., ill.

Dodson, E.O., 1960. Evolution: process and product. Reinhold, New York, 352 pp.

Doty, M.S. (ed.), 1961. Primary productivity measurement, marine and freshwater. *In* Proc. Conf. at Univ. Hawaii. Atomic Energy Comm., Div. Tech. Information, no. TID-7633, 237 pp.

Dowdeswell, W.H., R.A. Fisher and E.B. Ford, 1940. The quantitative study of populations in the Lepidoptera, I. *Polyommatus icarus* Rott. Ann. Eugenics, 10: 123–136.

Dudziak, J., 1951. Experiments on the plasticity of instinct in *Phryganea obsoleta* McLachlan (Trichoptera). Bull. Internatl. Acad. Polonaise, III, series B II, 1950: 145–171.

Dunbar, R.W., 1958. The salivary gland chromosomes of two sibling species of black flies included in *Eusimulium aureum* Fries. Canadian Jour. Zool. (Ottawa), 36: 23–44.

Edmondson, W.T., 1960. Reproductive rates of rotifers in natural populations. Mem. Inst. Italian Idrobiol., 12: 21–77.

Edmondson, W.T., 1964. The rate of egg production by rotifers and copepods in natural populations as controlled by food and temperature. Verhandl. Internatl. Ver. Limnologie 1962, (2): 673–675.

Edwards, G.A., 1953. Respiratory mechanisms; respiratory metabolism. *In* Roeder, 1953: 55–146.

Ehrlich, P.R., 1961a. Intrinsic barriers to dispersal in the checkerspot butterfly. Science, 134: 108–109.

Ehrlich, P.R., 1961b. Studies of the population structure of the checkerspot butterfly, *Euphydryas editha,* 1960. Res. Rept. no. 1, Jasper Ridge Biol. Exp. Area, 2 pp.

Ehrlich, P.R., and S.E. Davidson, 1961. Techniques for capture-recapture studies of Lepidoptera populations. Jour. Lepidopterists Soc., 14: 227–229.

Ehrlich, P.R., and R.W. Holm, 1963. The process of evolution. McGraw-Hill, New York, 347 pp., ill.

Emerton, J.H., 1961. The common spiders of the United States. . . . Dover, New York, 227 pp., ill.

Englemann, M.D., 1961. The role of soil arthropods in the energetics of an old field community. Ecol. Monogr., 31: 221–238.

Eriksen, C.H., [1961]. Respiration and substrate as factors influencing the distribution of the burrowing mayflies *Ephemera simulans* and *Hexagenia limbata.* Univ. Michigan, Ph.D. dissertation, 113 pp.

Eriksen, C.H., 1963a. The relation of oxygen consumption to substrate particle size in two burrowing mayflies. Jour. Exp. Biol., 40: 447–453.

Eriksen, C.H., 1963b. Respiratory regulation in *Ephemera simulans* Walker and *Hexagenia limbata* (Serville) (Ephemeroptera). Jour. Exp. Biol., 40: 455–467.

Evans, F.C., and F.E. Smith, 1952. The intrinsic rate of natural increase for the human louse, *Pediculus humanus* L. Amer. Nat., 86: 299–310.

Fabre, J.H., 1876–1904. Souvenirs Entomologiques. Delegrave, Paris, 11 vols. (Many parts have been translated into English under various titles.)

Flint, O.S., 1956. The life history of the genus *Frenesia* (Trichoptera: Limnephilidae). Bull. Brooklyn Ent. Soc., 51: 93–108.

Fox, H.M., B.G. Simmonds and R. Washburn, 1935. Metabolic rates of ephemerid nymphs from swiftly flowing and from still waters. Jour. Exp. Biol., 12: 179–184.

Fox, H.M., and C.A. Wingfield, 1938. A portable apparatus for the determination of oxygen dissolved in a small volume of water. Jour. Exp. Biol., 15: 437–445.

Fox, H.M., C.A. Wingfield and B.G. Simmonds, 1933. Metabolic rates of aquatic arthropods from different habitats. Jour. Exp. Biol., 10: 67.

Fox, H.M., C.A. Wingfield and B.G. Simmonds, 1937. The oxygen consumption of ephemerid nymphs from flowing and from still waters in relation to the concentration of oxygen in the water. Jour. Exp. Biol., 14: 210–218.

Fox, J.W., 1963. A portable ultra-violet insect trap. Ann. Carnegie Mus., 36: 205–212.

Fox, R.M., and J.W. Fox, 1964. Introduction to comparative entomology. Reinhold, New York, 450 pp., ill.

Fraenkel, G., and D.L. Gunn, 1940. Orientation of animals. Clarenden Press, Oxford, England, 352 pp., ill. (Reprinted 1961 by Dover, New York)

Frank, P.W., 1957. Coactions in laboratory populations of two species of *Daphnia*. Ecology, 38: 510–519.

Frank, P.W., 1960. Prediction of population growth form in *Daphnia pulex* cultures. Amer. Nat., 94: 357–372.

Frank, P.W., C.D. Boll and R.W. Kelly, 1957. Vital statistics of laboratory cultures of *Daphnia pulex* De Geer as related to density. Physiol. Zool., 30: 287–305.

Frankhauser, G., and L.E. Reik, 1935. Experiments on the case building of the caddis fly larva *Ptilostomis postica* Walker. Physiol. Zool., 8: 337–359.

Frings, H., 1946. The mounting of living insects for observation and study. Turtox News, 24: 149–154.

Frings, M., and H. Frings, 1960. Sound production and sound reception by insects: a bibliography. Pennsylvania State Univ. Press, University Park, Pa., 108 pp.

Fry, F.E.J., 1958. Temperature compensation. Ann. Rev. Physiol., 20: 207–224.

Galtsoff, P.S., F.E. Lutz, P.S. Welch and J.G. Needham (eds.), 1937. Culture methods for invertebrate animals. A compendium prepared . . . Section F of the American Association for the Advancement of Science. Reprinted. Dover (1959), New York, 590 pp.

Gause, G.F., and A.A. Witt, 1935. Behavior of mixed population and the problem of natural selection. Amer. Nat., 69: 596–609.

Gertsch, W.J., 1949. American spiders. Van Nostrand, Princeton, N.J., 285 pp., ill.

Giese, A.A., 1962. Cell physiology. Saunders, Philadelphia, 592 pp., ill.

Gilbert, L.I., 1963. Hormones controlling reproduction and molting in invertebrates. Comp. Endocrinol. 2: 1–46.

Gilbert, L.I., and H.A. Schneiderman, 1960. The development of a bioassay for the juvenile hormone of insects. Trans. Amer. Micros. Soc., 79: 38–67.

Gilbert, L.I., and H.A. Schneiderman, 1961. Some biochemical aspects of insect metamorphosis. Amer. Zool., 1: 11–51.

Gilmour, D., 1961. The biochemistry of insects. Academic Press, New York, 343 pp.

Golley, F.B., 1961. Energy values of ecological materials. Ecology, 42: 581–584.

Golley, F.B., 1964. Table of calorific equivalents. (Mimeographed) Univ. Georgia, Athens, 7 pp.

Gorter, F.G., 1928. Experiments on the case building of a caddis worm (*Limnephilus flavicornis* Fabr.). Tijdschr. Nederlandsche Dierk. Ver., (1) 3: 90–93.

Gottlieb, F.J., 1963. A method for mounting *Drosophila* body parts. Drosophila Information Serv., 37: 140.

Gross, L., and I. Halpern, 1961. Immobilization of fruit flies by hypothermia. Turtox News, 39: 258–259.

Gunn, D.L., and J.S. Kennedy, 1936. Apparatus for investigating the reactions of land arthropods to humidity. Jour. Exp. Biol., 13: 450–459.

Gunn, D.L., and D.P. Pielou, 1940. The humidity behavior of the mealworm beetle *Tenebrio molitor* L., III. The mechanism of the reaction. Jour. Exp. Biol., 17: 307–316.

Hall, D.J., 1964. An experimental approach to the dynamics of a natural population of *Daphnia galeata mendotae*. Ecology, 45: 94–110.

Haller, H.L., F. Acree and S.F. Potts, 1944. The nature of the sex attractant of the female gypsy moth. Jour. Amer. Chem. Soc., 66: 1659–1662.

Hanna, H.M., 1957. Observations on case building by the larvae of *Limnephilus politus* McLachlan and *L. marmoratus* Curtis (Trichoptera: Limnephilidae). Proc. Roy. Ent. Soc. London, (A) 32: 47–52.

Hanna, H.M., 1959. The growth of larvae and their cases and the life cycles of five species of caddis flies (Trichoptera). Proc. Roy. Ent. Soc. London, (A) 34: 121–129.

Hanna, H.M., 1960. Methods of case building and repair by larvae of caddis flies. Proc. Roy. Ent. Soc. London, (A) 35: 97–106.

Harker, J.E., 1961. Diurnal rhythms. Ann. Rev. Ent., 6: 131–146.

Haskell, P.T., 1961. Insect sounds. Witherby, London and Quadrangle Books, Chicago, 189 pp., ill.

Hazen, W.E., 1964. Readings in population and community ecology. Saunders, Philadelphia, 388 pp.

Heinrich, C., 1956. American moths of the subfamily Phycitinae. Bull. U.S. Natl. Mus., 207: 1–581.

Hodgson, E.S., 1956. Electrophysiological studies of arthropod chemoreception, I. General properties of the labellar chemoreceptors of Diptera. Jour. Cell Comp. Physiol., 48: 51–76.

Hodgson, E.S., 1958. Chemoreception in arthropods. Ann. Rev. Ent., 3: 19–36.

Huffaker, C.B., 1958. Experimental studies on predation: dispersion factors and predator-prey oscillations. Hilgardia, 27: 343–383.

Ivlev, W.W., 1939a. [Transformation of energy by aquatic animals.] Internatl. Rev. Hydrobiol., 38: 449–458.

Ivlev, W.W., 1939b. [Balance of energy in carps.] Zool. Zhurnal, 18: 303–318.

Jacobson, M., and M. Beroza, 1963. Chemical insect attractants. Science, 140: 1367–1373.

Jacobson, M., and M. Beroza, 1964. Insect attractants. Sci. Amer., 211 (2): 20–27.

Jacobson, M., M. Beroza and R.T. Yamamoto, 1963. Isolation and identification of the sex attractant of the American cockroach. Science, 139: 48–49.

Jacques, H.E., 1947. How to know the insects. Wm. C. Brown Co., Dubuque, Iowa, 205 pp., ill.

Jander, R., 1963. Insect orientation. Ann. Rev. Ent., 8: 95–114.

Jenkin, P.M., 1962. Animal hormones, a comparative survey. Part I, kinetic and metabolic hormones. Pergamon Press, New York, 310 pp.

Jenkins, D.W., 1957. Radioisotopes in entomology. *In* C.L. Comar (ed.), 1957, Atomic energy and agriculture: 195–229. Amer. Assoc. Adv. Sci., Publ. 49.

Jenkins, D.W., 1962. Radioisotopes in ecological and biological studies of agricultural insects. *In* Proc. of symposium on radioisotopes and radiation in entomology: 3–21. Internatl. Atomic Energy Agency, Vienna.

Johannsen, O.A., and F.H. Butt, 1941. Embryology of insects and myriapods. McGraw-Hill, New York, 462 pp.

Jones, J.C., 1962. Current concepts concerning insect hemocytes. Amer. Zool., 2: 209–246.

Juday, C., 1940. The annual energy budget of an inland lake. Ecology, 21: 438–450.

Kalmus, H., 1960. 101 simple experiments with insects. Doubleday, Garden City, N.Y., 194 pp.

Karlson, P., and M. Nachtigall, 1961. Ein biologischer Test zur quantitativen Bestimmung der Juvenilhormon-aktivität von Insektenextrakten. Jour. Insect Physiol., 7: 210–215.

Kaston, B.J., and E. Kaston, 1953. How to know the spiders. W. C. Brown, Dubuque, Iowa, 220 pp.

King, R.C., 1962. Genetics. Oxford Univ. Press, New York, 347 pp., ill.

Kloft, W., 1962. Technical problems of radioisotope measurement in insect metabolism. *In* Proc. symposium on radioisotopes and radiation in entomology: 163–172. Internatl. Atomic Energy Agency, Vienna.

Kloft, W., and P. Ehrhardt, 1962. Studies on the assimilation and excretion of labelled phosphate in aphids. *In* Proc. of symposium on radioisotopes and radiation in entomology: 181–190. Internatl. Atomic Energy Agency, Vienna.

Klopfer, P.H., 1962. Behavioral aspects of ecology. Prentice-Hall, Englewood Cliffs, N.J., 172 pp.

Krough, A., 1916. Respiratory exchange of animals and man. Longmans, Green, New York, 173 pp., ill.

Krough, A., 1935. Micro-Winkler method for the determination of dissolved oxygen. Analysis edition, Indust. Engin. Chem., 7: 131–135.

Krough, A., 1941. Comparative physiology of respiratory mechanisms. Univ. Pennsylvania Press, Philadelphia, 172 pp.

Lawson, F.R., C.R. Gentry and J.M. Stanley, 1963. Effect of light traps on hornworm populations in large areas. U.S. Dept. Agr., ARS-33-91: 1–19.

Legler, J.M., 1964. Tape recordings of frog calls. Turtox News, 42: 68–69.

Lewallen, L.L., 1957. Paper chromatography studies of the *Anopheles maculipennis* complex in California (Diptera: Culicidae). Ann. Ent. Soc. Amer., 50: 602–606.

Lincoln, F.C., 1930. Calculating waterfowl abundance on the basis of banding returns. U.S. Dept. Agr., Cir. 118: 1–4.

Lindeman, R.L., 1941. Seasonal food-cycle dynamics in a senescent lake. Amer. Midland Nat., 26: 636–673.

Lindeman, R.L., 1942. The trophic-dynamic aspect of ecology. Ecology, 23: 399–418.

Lloyd, J.T., 1921. The biology of North American caddis fly larvae. Lloyd Library, Ent. Ser. 1, Bull. 21: 1–124.

Lonert, A.C., 1964. Chromatography. Turtox News, 42: 50–53.

Lotka, A.J., 1945. Population analysis as a chapter in the mathematical theory of evolution. *In* Le Gros Clark, W.E., and P.B. Medawar (eds.), 1945. Essays on growth and form: 355–385. Oxford Univ. Press, London.

Lougee, L.B., 1963. The web of the spider. Cranbrook Inst. Sci., Bull. 46: 44 pp.

Lund, E.J., 1921. A micro-Winkler method for quantitative determination of dissolved oxygen. Proc. Soc. Exp. Biol. Med., 19: 63.

Lyman, F.E., 1945. Reactions of certain nymphs of *Stenoneura* (Ephemeroptera) to light as related to habitat preference. Ann. Ent. Soc. Amer., 38: 234–236.

Macfadyen, A., 1948. The meaning of productivity in biological systems. Jour. Animal Ecol., 17: 75–80.

Macfadyen, A., 1963. Animal ecology, Pitman, London, 344 pp.

Macan, T.T., 1963. Freshwater ecology. Wiley, New York, 338 pp.

Maddrek, S.H.P., 1963. Excretion in the blood-sucking bug *Rhodnius prolixus* Stäl, I. The control of diuresis. Jour. Exp. Biol., 40: 247–256.

Marshall, W.S., and C.T. Vorhies, 1905. The repair and rebuilding of the larval case of *Platyphylax designatus* Walker. Biol. Bull., 9: 232–244.

Martin, A., Jr., 1947. An introduction to the genetics of *Habrobracon juglandis* Ashmead. Hobson Press, New York, 205 pp., ill.

Mayr, E., 1947. Systematics and the origin of species. Columbia Univ. Press, New York, 334 pp., ill.

Mayr, E., 1958. Behavior and systematics. *In* Roe, A., and G.G. Simpson (eds.), Behavior and evolution: 341–362. Yale Univ. Press, New Haven, Conn.

McNabb, C.D., 1960. Enumeration of freshwater phytoplankton concentrated on the membrane filter. Limnology Oceanography, 5: 57–61.

Mecom, J.O., and K.W. Cummins, 1964. A preliminary study of the trophic relationships of the larvae of *Brachycentrus americanus* (Banks) (Trichoptera: Brachycentridae). Trans. Amer. Micros. Soc., 83: 233–243.

Merrill, D., [1964]. An analysis of case building behavior in four species of trichopteran larvae. Univ. Michigan, Ann Arbor, Ph.D. dissertation, 166 pp. Jour. Exp. Biol. [In press.]

Metcalf, Z.P., and C.L. Metcalf, 1928. A key to the principal orders and families of insects (3rd ed. rev., ill.). Privately printed, 23 pp.

Michener, C.D., 1952. The Saturniidae (Lepidoptera) of the western hemisphere. Bull. Amer. Mus. Nat. Hist., 98: 235–502.

Michigan, Univ. of, School Public Health, 1957. Radioactive liquid wastes. Univ. Michigan Publ. Distributions, Ann Arbor, Mich., 90 pp.

Micks, D.W., 1954. Paper chromatography as a tool for mosquito taxonomy: the *Culex pipiens* complex. Nature, 174: 217–221.

Miller, R.S., 1964. Larval competition in *Drosophila melanogaster* and *D. simulans.* Ecology, 45: 132–148.

Millipore Filter Corp., 1963. Application data manual for microbiological analysis of water and milk. Publ. ADM-10, Millipore Filter Corp., Bedford, Mass., 20 pp.

Milne, D.J., 1943. The distribution and life histories of the caddis flies of Wakesiu Lake, Saskatchewan. Canadian Ent., 75: 191–198.

Minnich, D.E., 1926. The chemical sensitivity of the tarsi of certain muscid flies. Biol. Bull., 51: 166–178.

Moore, J.A., 1952. Competition between *Drosophila melanogaster* and *Drosophila simulans,* I. Population cage experiments. Evolution, 6: 407–420.

Morris, R.W., 1963. A modified Barcroft respirometer for study of aquatic animals. Turtox News, 41: 22–23.

Munson, S.C., 1953. The hemocytes, pericardial cells and fat body. *In* Roeder, 1953: 218–231.

Murphy, H., 1919. Observations on the egg-laying of the caddis fly *Brachycentrus nigrisoma* Banks, and on the habits of young larvae. Jour. N.Y. Ent. Soc., 27: 154–159.

National Academy of Sciences, 1958. Handbook of respiration. Saunders, Philadelphia, 403 pp.

Naumann, E., 1925. Handbuch der biologischen Arbeitsmethoden. Abt. 9, Methoden zur Erforschung der Leistungen des tierschen Organismus, Tiel 2, Halften 1–3: 543–652.

Needham, J.G., and P.R. Needham, 1962. A guide to the study of fresh-water biology. Holden-Day, San Francisco, Calif., 108 pp.

Newman, D.W., 1964. Instrumental methods of experimental biology. Macmillan, New York, 592 pp.

Nielsen, A., 1948. Postembryonic development and biology of Hydroptilidae. Biol. Skr., 5: 1–200.

Noyes, A.A., 1914. The biology of the net-spinning Trichoptera of Cascadilla Creek. Ann. Ent. Soc. Amer., 7: 251–275.

Nuclear-Chicago Corp., 1959. Operations and maintenance manual for model GW1 carbon-14 glassware system. Nuclear-Chicago Corp., 333 East Howard Ave., Des Plaines, Ill., manual BK-092, 8 pp.

Nuclear-Chicago Corp., 1961. Installation and operation manual for model 6000 Dynacon electrometer system. Nuclear-Chicago Corp. manual BK-122, 32 pp.

O'Brien, R.D., and L.S. Wolfe, 1964. Radiation, radioactivity and insects. Academic Press, New York, 211 pp.

Odum, E.P., 1959. Fundamentals of ecology. Saunders, Philadelphia, 546 pp.

Odum, E.P., 1963. Ecology. Holt, Rinehart & Winston, New York, 152 pp.

Pant, R., and H.C. Agrawal, 1964. Free amino acids of the hemolymph of some insects. Jour. Insect Physiol., 10: 443–446.

Park, T., 1948. Experimental studies of interspecies competition, I. Competition between populations of the flour beetles *Tribolium confusum* Duval and *Tribolium castaneum* Herbst. Ecol. Monogr., 18: 265–308.

Park, T., 1954. Experimental studies of interspecies competition, II. Temperature, humidity and competition in two species of *Tribolium*. Physiol. Zool., 27: 177–238.

Park, T., 1955. Experimental competition in beetles, with some general implications. *In* Cragg, J.B., and N.W. Pirie, 1955. The numbers of man and animals, 152 pp., Oliver & Boyd, London.

Park, T., 1962. Beetles, competition and populations. Science, 138: 1369–1375.

Park, T., E.V. Gregg and C.Z. Lutherman, 1941. Studies in population physiology, X. Interspecific competition in populations of granary beetles. Physiol. Zool., 14: 395–430.

Parr Instrument Co., 1958. Instructions for the 1411 combustion calorimeter. Parr Instrument Co., Moline, Ill., manual 128, 22 pp.

Parr Instrument Co., 1960. Oxygen bomb calorimetry and combustion methods. Parr Instrument Co., manual 130, 56 pp.

Patton, R.L., 1963. Introductory insect physiology. Saunders, Philadelphia, 245 pp.

Pearce, E.M., 1956. Rapid determination of radio carbon in animal tissues. Analyt. Chem., 28: 1762–1765.

Pearl, R., T. Park and J.R. Miner, 1941. Experimental studies on the duration of life, XVI. Life tables for the flour beetle *Tribolium confusum* Duval. Amer. Nat., 75: 5–19.

Pearson, O.P., 1948. Metabolism and bioenergetics. Sci. Monthly, 66: 131–134.

Pennak, R.W., 1952. Fresh-water invertebrates of the United States. Ronald Press, New York, 769 pp.

Perfiliev, B.W., 1929. Zur Mikrobiologie der Bodenablagerungen. Verh. Internl. Ver. Limnolog., 4: 107–143.

Peterson, A., 1934–1937. A manual of entomological equipment and methods. Privately printed by the author. Part 1 (1934), [182 pp.], part II (1937), 334 pp.

Pfadt, R.E. (ed.), 1962. Fundamentals of applied entomology. Macmillan, New York, 668 pp.

Phillipson, J., 1960a. A contribution to the feeding biology of *Mitopus morio* (F.) (Phalangida). Jour. Animal Ecol., 29: 35–43.

Phillipson, J., 1960b. The food consumption of different instars of *Mitopus morio* (F.) (Phalangida) under natural conditions. Jour. Animal Ecol., 29: 299–307.

Picker X-Ray Corp., 1960. Radioisotope training manual. Part I, Theory. Picker X-Ray Corp., White Plains, N.Y., 62 pp.

Pielou, D.P., 1940. The humidity behavior of the mealworm beetle *Tenebrio molitor* L., II. The humidity receptors. Jour. Exp. Biol., 17: 295–306.

Pielou, D.P., and D.L. Gunn, 1940. The humidity behavior of the mealworm *Tenebrio molitor* L., I. The reaction to differences in humidity. Jour. Exp. Biol., 17: 286–294.

Pierce, G.W., 1949. The songs of insects. . . . Harvard Univ. Press, Cambridge, Mass., 329 pp.

Prosser, C.L., D.W. Bishop, F.A. Brown, T.L. Jahn and V.J. Wulff, 1952. Comparative animal physiology. Saunders, Philadelphia, 888 pp.

Prosser, C.L., and F.A. Brown, 1961. Comparative animal physiology (2nd ed.), Saunders, Philadelphia, 688 pp.

Ralph, C.L., 1962. Laboratory manual in cell physiology. Publ. by author, Univ. Pittsburgh, Pittsburgh, Pa., 62 pp.

Rehn, J.A.G., and H.J. Grant, 1961. A monograph of the Orthoptera of North America (north of Mexico), vol. I. Monogr. Acad. Nat. Sci. Philadelphia, 12: 1–255, ill.

Richman, S., 1958. The transformation of energy by *Daphnia pulex*. Ecol. Monogr., 28: 273–291.

Rizki, T.M., 1962. Experimental analysis of hemocyte morphology in insects. Amer. Zool., 2: 247–256.

Roeder, K.D. (ed.), 1953. Insect physiology. Wiley, New York, 1100 pp.

Roeder, K.D., 1955. Spontaneous activity and behavior. Sci. Monthly, 80: 362–370.

Roeder, K.D., 1963. Nerve cells and insect behavior. Harvard Univ. Press, Cambridge, Mass., 188 pp.

Roeder, K.D., and S. Roeder, 1939. Electrical activity in the isolated ventral nerve cord of the cockroach, I. The action of pilocarpine, nicotine, eserine and acetyl choline. Jour. Cellular Comp. Physiol., 14: 1–12.

Ross, H.H., 1938. Descriptions of nearctic caddis flies. Bull. Illinois Nat. Hist. Surv., 21: 101–183.

Ross, H.H., 1959. Trichoptera. *In* Edmondson, W.T. (ed.), 1959. Freshwater biology: 1024–1049. Wiley, New York.

Ross, H.H., 1944. The caddis flies or Trichoptera of Illinois. Bull. Illinois Nat. Hist. Surv., 23: 1–326.

Ross, H.H., 1956. Evolution and classification of the mountain caddis flies. Univ. Illinois Press, Urbana, Ill., 213 pp.

Ross, H.H., 1962. A synthesis of evolutionary theory. Prentice-Hall, Englewood Cliffs, N.J., 387 pp.

Roth, L.M., and E.R. Willis, 1951a. Hygroreceptors in adults of *Tribolium*. Jour. Exp. Zool., 116: 527–570.

Roth, L.M., and E.R. Willis, 1951b. Hygroreceptors in Coleoptera. Jour. Exp. Zool., 117: 451–488.

Rothfels, K.H., 1958. Salivary gland chromosomes and phylogenetic interrelations in black flies. Proc. 10th Internatl. Congr. Ent. (Montreal, 1956), 2: 907.

Rothfels, K.H., and R.W. Dunbar, 1953. The salivary gland chromosomes of the black fly *Simulium vittatum* Zett. Canadian Jour. Zool. (Ottawa), 31: 226–241.

Rothschild, W., and K. Jordan, 1898–1899. A monograph of *Charaxes* and allied prionopterous genera. Novitates Zool., 5: 545–601; 6: 220–286.

Sarkaria, D.S., S. Battini and R.L. Patton, 1951. A rapid method for clinical study of cockroach blood cells. Canadian Ent., 83: 329–332.

Schneiderman, H.A., and L.I. Gilbert, 1964. Control of growth and development in insects. Science, 143: 325–333.

Schnierla, T.C., 1953. [Insect behavior.] Chapters 25–28. *In* Roeder, 1953: 656–779.

Scholander, P.F., 1952. Microvolumetric respirometry. Jour. Gen. Physiol., 35: 375–395.

Scholander, P.F., and G.A. Edwards, 1942. Microrespiration apparatus. Rev. Sci. Instruments, 13: 292–295.

Sessions, J., [1961]. Some respiratory studies on Trichoptera from different habitats. Univ. Michigan, Ann Arbor, Mich., unpublished research report, 39 pp.

Shelford, V.E., 1913. Animal communities in temperate America. Univ. Chicago Press, Chicago, 362 pp.

Sibly, C.G., 1960. The electrophoretic patterns of avian egg-white proteins as taxonomic characters. Ibis, 102: 215–284.

Simpson, G.G., 1961. Principles of animal taxonomy. Columbia Univ. Press, New York, 247 pp., ill.

Sinnott, E.W., L.D. Dunn and T. Dobzhansky, 1958. Principles of genetics, 5th ed. McGraw-Hill, New York, 459 pp., ill.

Slobodkin, L.B., 1954. Population dynamics in *Daphnia otusa* Kurz. Ecol. Monogr., 24: 69–88.

Slobodkin, L.B., 1961. Growth and regulation of animal populations. Holt, Rinehart & Winston, New York, 184 pp.

Slobodkin, L.B., 1962. Energy in animal ecology. *In* Cragg, 1962: 69–101.

Slobodkin, L.B., and S. Richman, 1961. Calories/gm. in species of animals. Nature, 191: 299.

Slyke, D.D. Van, J. Plazin and J.R. Weisiger, 1951. Reagents for the Van Slyke-Folch wet carbon combustion. Jour. Biol. Chem., 191: 299–304.

Smirnov, N.N., 1962. On nutrition of caddis worms *Phryganea grandis* L. Hydrobiologia, 19: 252-261.

Smith, F.E., 1962. Experimental methods in population dynamics: a critique. Ecology, 33: 441–450.

Smith, N.A., [1964]. Neuroendocrine control of heart function in the cockroach. Univ. Pittsburgh, Pittsburgh., Ph.D. dissertation in preparation.

Snodgrass, R.E., 1935. Principles of insect morphology. McGraw-Hill, New York, 667 pp.

Spieth, H.T., 1952. Mating behavior within the genus *Drosophila* (Diptera). Bull. Amer. Mus. Nat. Hist., 99: 399–474.

Srb, A.M., and R.D. Owen, 1958. General genetics. Freeman, San Francisco, 561 pp., ill.

Strickberger, M.W., 1962. Experiments in genetics with Drosophila. Wiley, New York, 144 pp., ill.

Strickland, J.D.H., 1960. Measuring the production of marine phytoplankton. Bull. Fisheries Res. Board, Canada, 122: 1–172.

Sweadner, W.R., 1937. Hybridization and the phylogeny of the genus *Platysamia*. Ann. Carnegie Mus., 25: 163–242.

Thienemann, A., 1926. Der Nahrungskreislauf im Wasser. Verhandl. Deutsch. Zool. Gesell., 31 [Zool. Anz. Suppl., 2]: 29–79.

Thienemann, A., 1931. Der Produktionsbegriff in der Biologie. Arch. Hydrobiol., 22: 616.

Tietz, H.M., 1963. Illustrated keys to the families of North American insects. Burgess, Minneapolis, 206 pp.

Tindall, A.R., 1960. The larval case of *Triaenodes bicolor* Curtis (Trichoptera: Leptoceridae). Proc. Roy. Ent. Soc. London, (A) 35: 93–96.

Tizard, H.T., 1932. Thermochemistry. Encyclopaedia Britannica (14th ed.), 22: 83–91.

Tobias, J.M., 1943. Microrespiration techniques. Physiol. Rev., 23: 51–75.

Trama, F.B., [1957]. The transformation of energy by an aquatic herbivore, *Stenonoma pulchellum* (Ephemeroptera). Univ. Michigan, Ann Arbor, Mich., Ph.D. dissertation.

Tuxen, S.L. (ed.), 1956. Taxonomist's glossary of genitalia of insects. E. Munksgaard, Copenhagen, Denmark, 284 pp., ill.

Umbreit, W.W., R. Burris and J. Stauffer, 1957. Manometric techniques. Burgess, Minneapolis, 203 pp.

Usinger, R.L. (ed.), 1956. Aquatic insects of California, with keys to North American genera and California species. Univ. California Press, Berkeley, 508 pp.

Verduin, J., 1951. Photosynthesis in naturally reared aquatic communities. Plant Physiol., 26: 45–49.

Vorhies, C.T., 1905. Habits and anatomy of the larva of the caddis fly *Platyphylax designatus* Walker. Trans. Wisconsin Acad. Sci. Arts Lett., 15: 108–123.

Vorhies, C.T., 1909. Studies on the Trichoptera of Wisconsin. Trans. Wisconsin Acad. Sci. Arts Lett., 16: 647–739.

Waldron, I., 1964. Courtship sound production in two sympatric sibling *Drosophila* species. Science, 144: 191–193.

Webster, D.A., and P.C. Webster, 1943. Influence of water current on case weight in larvae of the caddis fly *Goera calcarata* Banks. Canadian Ent., 75: 105–108.

Welsh, J.H., and R.I. Smith, 1960. Laboratory exercises in invertebrate physiology. Burgess, Minneapolis, 179 pp.

Welsh, P.S., 1948. Limnological methods. Blakiston, Philadelphia, 381 pp.

Wessenberg-Lund, C., 1943. Biologie der Süsswasserinsekten. Springer, Berlin, 682 pp.

Whiting, P.W., 1918. Sex determination and biology of a parasitic wasp, *Habrobracon brevicornis* (Wesmael). Biol. Bull., 34: 250–256.

Whitten, J.M., 1964. Giant polytene chromosomes in hypodermal cells of developing footpads of dipteran pupae. Science, 143: 1437–1438.

Wiggins, G.B., 1954. The caddis fly genus *Beraea* in North America. Contrib. Roy. Ontario Mus. Zool. Paleont., 39: 1–18.

Wiggins, G.B., 1956. The Kitamaiidae, a family of caddis flies new to North America. Contrib. Roy. Ontario Mus. Zool. Paleont., 44: 1–10.

Wiggins, G.B., 1960. A preliminary systematic study of North American larvae of the caddis fly family Phryganeidae (Trichoptera). Canadian Jour. Zool. (Ottawa), 38: 1153–1170.

Wiggins, G.B., 1962. A new subfamily of phryganeid caddis flies from western North America (Trichoptera: Phryganeidae). Canadian Jour. Zool. (Ottawa), 40: 879–891.

Wigglesworth, V.B., 1956. Principles of insect physiology (5th ed.). Methuen, London and Wiley, New York, 130 pp., ill.

Wigglesworth, V.B., 1958. Some methods for assaying extracts of the juvenile hormone in insects. Jour. Insect Physiol., 2: 73–84.

Wigglesworth, V.B., 1959. The control of growth and form. Cornell Univ. Press, Ithaca, N.Y., 140 pp.

Wigglesworth, V.B., 1963. The juvenile hormone effect of farnesol and some related compounds: quantitative experiments. Jour. Insect Physiol., 9: 105–119.

Wilde, J. de, 1962. Photoperiodism in insects and mites. Ann. Rev. Ent., 7: 1–26.

Williams, C.M., 1956. The juvenile hormone of insects. Nature, 178: 212–213.

Williams, C.M., 1958. The juvenile hormone. Sci. Amer., 198: 67–74.

Wintringham, F.P.W., 1959. An electrolytic respirometer. Lab. Practice, 8: 372–375.

Wittig, G., 1962. The pathology of insect blood cells: a review. Amer. Zool., 2: 257–273.

Zimmering, S., 1948. Competition between *Drosophila pseudoobscura* and *Drosophila melanogaster* in population cages. Amer. Nat., 82: 326–330.

INDEX

abdomen, 2, 7–8, 15, 22–23
Actias, 60, 123
Anagasta, 57, 144, 147–148
anal plate, 14, 15
antenna, 3, 11, 13, 15, 20
  aristate, 20
  clavate, 20
  doubly bipectinate, 20
  filiform, 20
  lamellate, 20
  plumose, 20
Antheraea polyphemus, 97, 100, 123
anus, 12
Apis, 21
apodemes, 8, 9
  dissection of, 8
  thoracic, 9
Arachnida, 23, 113–116
  gnathochilarium, 23
  morphology, 23
  orientation, 113–116
  web building, 113–116
  web collection method, 113
  arista, 14, 20
attraction, sexual, 60–62, 122–123
  Actias selene, 60
  chemical basis, 60–61
  Hyalophora cecropia, 60–61
  Hyalophora promethea, 60–61
  Lymantria dispar, 60

Blaberus, 2–9, 77, 94, 144
  morphology, 2–9
  nervous system, 94
  respiration, 77
  sexual distinction, 7
blood, 71–74
  cells, see hemocytes
bristles, 14
bursa copulatrix, 19
butterfly culture, 146

cardo, 4, 5
carina, 34
caudal prolegs, 9
cercus, 7, 8, 22
  median filament, 22
cervix, 2
chemical senses, 59, 60–62, 122–123
  common, 59
  gustatory, 59
  olfactory, 59
Chi square test, 51
Chironomidae, 69, 70
chromatography, 48–49, 72–74
  apparatus, 72–74
  ratio of fronts (R_f), 74
chromosome, polytene, 54
clavola, see flagellum
clypeus, 3, 11
cockroach culture, 146
Colias, 118–120, 145
community energetics, 141–142
  calorimetry, 141–142
  energy budget, 142
condyle, 3, 4, 5, 7
  anterior, 3, 5
  anterior trochantral, 7
  posterior, 4, 5
  posterior trochantral, 7
cornutus, 18
corpora allata, 95–97
  secretions, 95
corpora cardiaca, 94–96
corporotentorium, see tentorium bridge
coxa, 5, 6, 11

coxopodite, 4
coxotrochantral joint, 5
Culicidae, 69–70
culture methods, 143–149
  butterflies, 146
  cockroaches, 146
  Drosophila, 147
  Gryllus, 147
  Habrobracon and *Anagasta,* 147–148
  Saturniidae, 148
  spiders, 148–149
  Tenebrio, 149
  Tribolium, 149
  Trichopteran larvae, 149

Danaus, 16, 22, 144
digestive system, 12–13
  crop, 13
  esophagus, 12, 13
  foregut, 13
  hindgut, 12, 13
  midgut, 12, 13
  pharynx, 12, 13
  proventriculus, 12, 13
Drosophila, 13–15, 50–56, 144, 145
  abdomen, 14–15
  bristles, 14
  competition, 130–131
  compound eye, 13
  courtship behavior, 110–111
  culture, 147
  genetic crosses, 50–54
  head, 13–14
  mating behavior, 46
  morphology, 13–15
  population census, 127–128
  respiration, 77
  salivary glands, 54–56
  sex linkage, 53–54
  sexual distinction, 14
  thorax, 14
  vestigial wing mutant, 52–53

ecdyson, 95–97
  action on dipteran chromosomes, 95–97
egg exit duct, 19
ejaculatory duct, 18
elytron, 15, 21
energetics, see community energetics
epicranium, 11
epimeron, 6
episternum, 6
equipment suppliers, 143, 150–151, 152–153
esophageal ganglion, see corpora cardiaca
exercises, species used in, 143–145
eye, 3, 11, 13
  compound, 3, 11, 13
  facet, 3
  ommatidium, 3
  simple, 11

fastigium, 34
femur, 7, 11
fenestra, 3
flagellum, 3
food plant selection, 63–67
  all-or-none response, 67
  Danaus plexippus, 64–65
  detection of odors, 64–65
  Papilio asterius, 65
  sensory centers, 65–67
  volatile oils, 65
food studies, see trophic dynamics
food web, 132
foramen, 18
  antennal, 3
foreleg, 5
frons, 3, 11
fronto-clypeus, 3

galea, 4, 5, 11, 21
ganglion, 12–13, 92–94
  spontaneous electrical discharge, 92–94
  subesophageal, 12
  supraesophageal, 12, 13
  thoracic, 12
gena, 3
genital arches, 15
genital capsule, 17
genitalia, 17–19
  female, 18–19
  male, 18
  preparation procedure, 17, 18, 151
genetic traits, hemizygous, 53
  heterozygous, 53
  homozygous, 53
gill, 11, 12, 22
  anal, 22
ginglymus, see condyle, anterior

glossa, 4, 5
gnathos, 18
Gryllus, 44–48, 111–113, 144, 145, 147
  calling songs, 44–48
  courtship communication, 111–113
  culture, 147
  species characteristics, 45
  stridulation structure, 111–112

habitats, lentic, 81
  lotic, 81
Habrobracon, culture, 57–58, 144, 147–148
  morphology, 57
  sex linkage, 56–58
halteres, 14
head, 2–4, 8–9, 11, 15, 20–21
heart rate, of insects, 89–90
hemelytron, 21
hemocytes, 71
hemolymph, 71
hook, accessory, 11
  anal, 9, 11
hormone, 95–100, 150
  brain, 95–97
  juvenile, see neotenin
  prothoracic gland, see ecdyson
humidity- and temperature-measuring devices, 67–69
humidity perception, see temperature
hump, spacing, 11
Hyalophora cecropia, 95–98, 122, 145
hypopharynx, 4–5

identification, behavioral, 35–48
  morphologic, 29–34
  physiologic, 48–49
insect songs, 44–46, 101, 111–113
intrinsic rate of natural increase (biotic potential), 128

Johnston's organ, 3
juvenile hormone, see neotenin

key, 26–44, 154
  arachnid orders, 29–30
  arthropod classes, 26–30
  insect orders, 29–33
  myriopod orders, 29–30
  Romaleinae, 33–35
  Trichoptera by case, 35–44

labium, 4, 5, 11, 15
labrum, 3, 11, 15
  medial notch, 11
lacinia, 4, 5
lacto-aceto-orcein solution, 55, 150
legs, 11, 22
  corbicula, 22
  modifications, 22
  stridulation, 22
Lepidoptera, 16–19
  female genitalia, 18–19
  male genitalia, 18
life table, 124
light orientation, 68
ligula, 4
Limnephilis, 9
lip, see labium

malpighian tubules, 12, 90–91
mandible, 3, 5, 11, 15
maxilla, 4, 5, 11, 15
  second, see labium
media and formulation, 150–152
mesonotum, 14–15
mesopleuron, 14
mesothorax, 4
metanotum, 14–15
metathorax, 4
Millipore filter, 80, 133–134, 153
Molanna uniophila, 89
molting, 95–100
mountants, 19, 71, 150–151
mouth, 4, 12
mouthparts, 3–4, 14
  gula, 20
  hypognathous, 3, 15, 20
  opisthognathous, 20
  proboscis, 21
  prognathous, 20
Musca, 62, 144
Myriapoda, 23, 26–28, 30

neotenin, 95–100, 150
  analogues, 100
  bioassay, 98–100
  extraction, 97–98
nerve chamber, 93, 153

nervous system, 12–13, 91–94
  ganglion, 12
  metameric independence, 91
  nerves, 12
  physiology, 91–94
  spontaneous electrical discharge, 92–94
  ventral nerve cord, 13
neurosecretory cells, 94, 95, 96
ninhydrin spray, 72, 74, 151
notum, 4

ocellus, 3, 13
occiput, 3
ovaries, 19
ovipositor, 14
oxygen bomb calorimetry, 141, 153
oxygen consumption, 75–85
  aquatic insects, 78–85
  temperature effects, 89
  terrestrial insects, 75–78
oxygen electrode instruments, 78

palpus, labial, 4, 5, 11, 21
  maxillary, 4, 5, 11, 13
papilla anala, 18
Papilio, 65, 144
paraglossa, 4, 5
pars intercerebralis, 94–95
parthenogenesis, 57
pedicel, 3
penis, 15, 17
Periplaneta, heart rate, 90
  respiration, 77
phenotype, 54
Phormia, 62, 144
photoreception, 70
Phryganea, 103–104, 107
Platycentropus, 9
pleurum, 14
podical plates, 8
population analysis, 117–131
  chemoattraction, 122–123
  competition, 129–131
  field census, 118–122
  laboratory census, 124–129
  Lindoln index, 121
  ultraviolet light trap, 123–124
postlabium, see postmentum
postmentum, 4, 5
postocciput, 3
post-tarsus, 5, 7, 11
pre-episternum, 6
prelabium, 4
prementum, 4, 5
primaries, see wings, forewings
Pringle's solution (saline), 55, 88, 90, 93
proboscis, 14, 21
pronotal disk, 34
pronotal shield, 2, 9, 11
pronotum, 14, 15
prothorax, 4
Pseudaletia, 123
pteropleuron, see mesopleuron
Ptilostomis, 9, 10, 12, 103, 104, 107
pulvillus, 7
Pycnopsyche guttifer, 9, 83–85, 86, 89, 104, 109
Pycnopsyche lepida, 83–85, 86, 89, 104–109

Q_{10}, 88–91

reception, sensory, 59–70, 92
respiration, 74–88
  buffering capacity of medium, 86–87
  energy equivalent, 86
  maintenance cost, 86
  manometric devices, 74–75
  respirometers, 75–76, 78–80
  resting rate, 83
  Warburg apparatus, 74–75
  Winkler method, 80–81
respiratory quotient (RQ), 74–75, 85–86
respiratory siphon, 22
Rhodnius, 92, 99

saccus, 18
Samia, 123
Sarcophaga, 62, 144
Sarkarig's stain, 71, 151
Saturniidae, 144, 148
  culture, 148
scape, 3
sclerite, 2–4, 6, 14
  antennal, 3
  axillary, 6, 14
  cervical, 2
  pleural, 4
scutellum, 14
secondaries, see wings, hindwings

sensillum, 11
setae, sensory, 11
sex comb, 14
shaft, see flagellum
shield, pronotal, 4
silk gland, 11, 12
silk orifice, 11, 12
sound production, 44–48, 111–113
  audiospectrographs, 47
  Gryllus, 44–48
special equipment, 152–153
spiders, 23–24, 26–30, 113–116, 145, 148, 149
  culture, 148–149
spine, tibial, 5, 7
spinning glands, see silk glands
spiracle, 6, 14
spur, tibial, 5, 7
sternellum, 6
sternite, 7, 14, 15
stria, 15, 16
sternum, 6, 7, 14
stipes, 4, 5
stylus, 7, 8
submentum, 11
suture, 3, 5, 6, 11
  antennal, 3
  coxal, 5, 6
  epicranial, 3, 11
  epistomal, 3
  fronto-clypeal, 3
  genal, 3
  occipital, 3
  parietal, 3
  pleural, 6
  postoccipital, 3
  subgenal, 3

tactile reception, 92
tap water conditioning, 86, 151
tarsus, 5, 7, 11
taste responses, 62–67
  Drosophila, 62
  Musca, 62
  negative, 62–63
  Phormia, 62
  positive, 62–63
  Sarcophaga, 62
  subliminal, 63
  threshold concentration, 63
tegmina, 2, 6
tegumen, 18
telopodite, 4
temperature and humidity, perception, 67–70
  Culex pipiens, 69–70
  gradient chamber, 67, 68
  hygroreceptors, 69
  Tendipes tentans, 69–70
  Tribolium, 68–70
temperature coefficient, see Q_{10}
Tenebrio, 77, 98–100, 145, 149
  culture, 149
  respiration, 77
tentorium, 8
  anterior arms, 8
  anterior pits, 8
  bridge, 9
  dorsal arms, 8
  posterior arms, 9
tergite, 4, 7, 11, 14, 15
tergum, 7
test cross, 53
thorax, 2, 4–7, 21–22
tibia, 5, 7, 11
tracer equipment, 137, 153
trachea, 12
  dorsal cephalic, 12
  longitudinal, 12
  ventral cephalic, 12
Tribolium, 15–16, 68–70, 77, 129–130, 144
  competition, 129–130
  culture, 149
  morphology, 15
  population census, 126–127
  respiration, 77
  responses, 68–70
  sexual distinction, 16
  species distinction, 15–16
trochanter, 4, 5, 6, 11
trophic dynamics (see also food plant selection), 131–141
  assimilation rate, 139
  calories, 132
  chromatographic technique, 135
  egestion rate, 139–141
  fluid ingestion, 135
  ingestion rate, 137–139
  Millipore filter technique, 133–135

trophic dynamics (*cont.*)
  precipitin test, 135
  solid ingestion, 133–135
  tracer techniques, 135–141
Trichoptera, 9–13, 35–44, 101–110
  behavior, 35–44, 101–110
  case building, *frontispiece,* 101–110
  case materials, 104–108
  culture methods, 103
  evolution of case building, 102
  external morphology, 9–11
  identification by case type, 35–44
  internal morphology, 11–13
  ontogeny of case building, 108–110

uncus, 18

vaginal plate, 14, 18
valve, 17, 18
veins, see wings
vertex, 3, 13
vinculum, 17, 18

whole mounts, 16–18, 151
wings, 2, 6, 14
  forewings, 2, 6, 14
  hindwings, 6, 14
  veins, 14
Wrights blood stain, 71